Rapid Prototyping Technology

MECHANICAL ENGINEERING
A Series of Textbooks and Reference Books

Founding Editor

L. L. Faulkner

Columbus Division, Battelle Memorial Institute
and Department of Mechanical Engineering
The Ohio State University
Columbus, Ohio

1. *Spring Designer's Handbook*, Harold Carlson
2. *Computer-Aided Graphics and Design*, Daniel L. Ryan
3. *Lubrication Fundamentals*, J. George Wills
4. *Solar Engineering for Domestic Buildings*, William A. Himmelman
5. *Applied Engineering Mechanics: Statics and Dynamics*, G. Boothroyd and C. Poli
6. *Centrifugal Pump Clinic*, Igor J. Karassik
7. *Computer-Aided Kinetics for Machine Design*, Daniel L. Ryan
8. *Plastics Products Design Handbook, Part A: Materials and Components; Part B: Processes and Design for Processes*, edited by Edward Miller
9. *Turbomachinery: Basic Theory and Applications*, Earl Logan, Jr.
10. *Vibrations of Shells and Plates*, Werner Soedel
11. *Flat and Corrugated Diaphragm Design Handbook*, Mario Di Giovanni
12. *Practical Stress Analysis in Engineering Design*, Alexander Blake
13. *An Introduction to the Design and Behavior of Bolted Joints*, John H. Bickford
14. *Optimal Engineering Design: Principles and Applications*, James N. Siddall
15. *Spring Manufacturing Handbook*, Harold Carlson
16. *Industrial Noise Control: Fundamentals and Applications*, edited by Lewis H. Bell
17. *Gears and Their Vibration: A Basic Approach to Understanding Gear Noise*, J. Derek Smith
18. *Chains for Power Transmission and Material Handling: Design and Applications Handbook*, American Chain Association
19. *Corrosion and Corrosion Protection Handbook*, edited by Philip A. Schweitzer
20. *Gear Drive Systems: Design and Application*, Peter Lynwander
21. *Controlling In-Plant Airborne Contaminants: Systems Design and Calculations*, John D. Constance
22. *CAD/CAM Systems Planning and Implementation*, Charles S. Knox
23. *Probabilistic Engineering Design: Principles and Applications*, James N. Siddall
24. *Traction Drives: Selection and Application*, Frederick W. Heilich III and Eugene E. Shube
25. *Finite Element Methods: An Introduction*, Ronald L. Huston and Chris E. Passerello

Additional Volumes in Preparation

Mechanical Engineering Software

Rapid Prototyping Technology

Selection and Application

Kenneth G. Cooper

National Aeronautics and Space Administration (NASA)
Marshall Space and Flight Center
Huntsville, Alabama

CRC Press
Taylor & Francis Group
Boca Raton London New York

CRC Press is an imprint of the
Taylor & Francis Group, an **informa** business

CRC Press
Taylor & Francis Group
6000 Broken Sound Parkway NW, Suite 300
Boca Raton, FL 33487-2742

© 2001 by Taylor & Francis Group, LLC
CRC Press is an imprint of Taylor & Francis Group, an Informa business

No claim to original U.S. Government works

ISBN-13: 978-0-8247-0261-8 (hbk)
ISBN-13: 978-0-367-39765-4 (pbk)

Visit the Taylor & Francis Web site at
http://www.taylorandfrancis.com

and the CRC Press Web site at
http://www.crcpress.com

For my loving wife, Jennifer. Thank you for your support and patience while I finished this writing.

Preface

Well, first off, thanks for buying this book! And, if you are only considering buying this book, or at least took the time to pick it up and read the preface, hey, thank you, too! Regardless, at this point you are hoping I don't ramble through the whole text like this, so let's get down to business, shall we?

This book was written by a technical professional who, for over seven years, has personally operated, modified, tweaked, repaired, and advanced at least one of every domestic rapid prototyping system on today's market. Just what is rapid prototyping? Well, here it refers to a collective set of manufacturing technologies that build parts in additive fashion, by growing slices of a material from the bottom to the top of the part, directly from computer-driven data.

My mechanical and materials engineering background has given me the desire to take the machines apart and see what makes them tick, and on most machines in this book I have done just that. I will not give out any trade secrets of any vendors or countries, only practical knowledge that, with a good bit of work, most anyone could probably find on their own, if only they had the time. Where I lacked expertise I tried to get information from more experienced folks in that niche, namely Curtis Manning for the ModelMaker chapter, Pat Salvail on the investment casting chapter and case study, and Anthony Springer for facilitating the wind-tunnel case study.

This text covers each rapid prototyping system at a fundamental level, that is, down to the basic operation of the machines along with typical uses and the pros/cons of each. The book covers not only the domestic systems that I have worked with, but also brief segments for foreign systems, as well as international progress in the rapid prototyping realm that I picked up with some savvy Internet surfing.

Finally, just in case I haven't completely covered something you are looking for, I've added a comprehensive recommended reading list of some of the best rapid prototyping texts on the market. So, happy reading, and I hope you find what you are looking for in these pages!

Kenneth G. Cooper, AST

Contents

1

What Is Rapid Prototyping?

1.1 Rapid Prototyping Defined

Rapid Prototyping (RP), as defined in this text, refers to the layer-by-layer fabrication of three-dimensional physical models directly from a computer-aided design (CAD). This additive manufacturing process provides designers and engineers the capability to literally print out their ideas in three dimensions. The RP processes provide a fast and inexpensive alternative for producing prototypes and functional models as compared to the conventional routes for part production.

The advantage of building a part in layers is that it allows you to build complex shapes that would be virtually impossible to machine, in addition to the more simple designs. RP can build intricate internal structures, parts inside of parts, and very thin-wall features just as easily as building a simple cube.

All of the RP processes construct objects by producing very thin cross sections of the part, one on top of the other, until the solid physical part is completed. This simplifies the intricate three-dimensional construction process in that essentially two-dimensional slices are being created and stacked together. For example, instead of trying to cut out a sphere with a detailed machining process, stacks of various-sized "circles" are built consecutively in the RP machine to create the sphere with ease.

RP also decreases the amount of operation time required by humans to build parts. The RP machines, once started, usually run unattended until the part is complete. This comes after the operator spends a small amount of time setting up the control program. Afterwards, some form of clean up operation is usually necessary, generally referred to as *post processing*. Nonetheless, the user intervention times still remain far less than that for traditional machining processes.

You can imagine the cost and time savings involved with such a process. Models can usually be built within hours, and the build materials for most processes are generally inexpensive. Some RP machines are small and environmentally friendly so that they can be placed directly into a designer's office, just like a common photocopy machine! Finally, having access to accurate, functional prototypes to verify concepts in the beginning phases of a project is an invaluable resource to any institution in the business of producing physical components.

1.2 Origins of Rapid Prototyping

RP stems from the ever-growing CAD industry, more specifically, the solid modeling side of CAD. *Solid modeling* is the branch of CAD that produces true three-dimensional objects in electronic format. A solid model has volume and is fully enclosed. It can be assigned materials properties such as mass and density, and the geometry data can be output in various formats to accommodate RP, stress analysis software packages, and, of course, computer numerically controlled machining (CNC).

Before solid modeling was introduced in the late 1980s, three-dimensional models were created with wireframes and surfaces. A *wireframe* is an approximate representation of a three-dimensional object, such as one would sketch with a pencil or on the chalkboard. Wireframes are sometimes deemed as two-and-a-half dimensional, for the fact that they only appear to be three-dimensional. Later on, the wireframes could actually have surfaces for visual enhancement and analysis. For example, a cube would be represented by six squares joined at the edges in three-dimensional coordinate space. But not until the development of true solid modeling could innovative processes such as RP be developed.

Charles Hull, who helped found 3D Systems in 1986, developed the first RP process. This process, called *stereolithography,* builds objects by curing thin consecutive slices of certain ultraviolet light-sensitive liquid resins with a low-power laser. This concept of layer additive construction has been capitalized on by various institutions in many different ways, which all have collectively been termed RP.

With the introduction of RP, CAD solid models could suddenly come to life. Designers and engineers now have the power to go through several iterations of a design in order to get the best possible performance for their needs.

1.3 The Design Process

In order to understand where RP fits into the manufacturing stream, it is best to begin with a description of a typical design process. Although the techniques for approaching a design-to-manufacture cycle may vary from business to business, a general path is taken by most of them for a mass-production item.

1   The Concept.   Any new product, or improvement of an old product, must start out as a concept, or idea. The source for such a concept may be based on a need, desire, or may simply be a random thought that entered someone's mind. However it came to be, in order to become a reality, the concept must be carried through the design-to-manufacture process.

2   Preliminary Design.   A preliminary design can range from a simple sketch on a napkin, to a two-dimensional drawing or even to a CAD solid model of the part needing to be built. The design can go through much iteration during this phase, as the designer determines the feasibility of the product through discussing with colleagues and co-workers and presenting to management, for instance. Other preliminary checks can now be performed with computers such as stress analysis, interference and fit, as well as visualization. RP can be useful in this phase by allowing the designer to have a physical representation to

help demonstrate the product's use and functions. This use of RP is referred to in this text as *concept verification.*

3 Preliminary Prototype Fabrication. Once a design has been given the go-ahead, a *prototype* must be fabricated to check out the design. Before RP, this phase of the design-to-manufacture process was carried out either by hand working or machining, both of which can be time consuming and expensive. Enter RP. Durable plastic or similar models can be fabricated quickly for *fit-check analysis* to determine if the design is the correct size, shape, etc. for the necessary application by demonstrating with a physical model. This step may repeat several times until the proper design is acquired. With the inexpensive RP alternative, it is now less of a burden to reiterate as opposed to before.

4 Short-run Production. Sometimes, a short-run production sequence may be necessary to further proof a part before entering into final production. Anywhere from ten to a few hundred parts may be manufactured and distributed for testing, verification, consumer satisfaction, etc. RP can be used in this phase for producing several small prototypes or, through a process often termed *rapid tooling,* be used to make several hundred parts inexpensively. It is crucial that any mistakes or flaws in the design are found during or before this stage, otherwise it could become very costly. For small runs of parts, this can even be the last phase of a program, without going to an expensive traditional tool at all.

5 Final Production. In this final step of the process, parts are typically either machined, injection molded, or cast in large numbers. The patterns for injections, or tooling, are usually machined from aluminum or steel so that they can be used several thousand, even hundreds of thousands, times. With the development of direct metal and ceramic processes, RP may yet reach this phase in the near future.

1.4 The Rapid Prototyping Cycle

As previously mentioned, RP can be useful in several steps of the design process. This segment describes the cycle from CAD design to the prototype part, which is actually embedded in the design process (Figure 1.1).

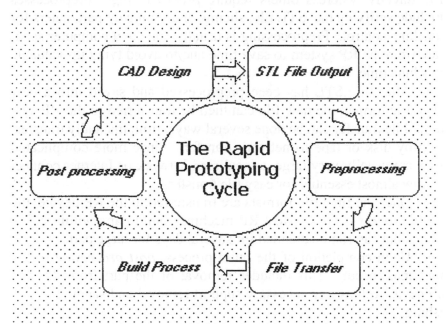

Figure 1.1 The RP cycle begins with the CAD design, and may be repeated inexpensively several times until a model of the desired characteristics is produced.

The first step is the CAD file creation. The final file or files must be in solid model format to allow for a successful prototype build. From the CAD file, an export format called the .STL file must be created.

The .STL file, so named by 3D Systems for STereoLithography, is currently the standard file format for all U.S. RP systems. STL files are triangulated representations of solid models. The individual triangles are represented by simple coordinates in a text file format. STL files are usually stored in binary format to conserve disk space.

After the .STL file is created, it must be prepared differently for various types of RP systems. Some systems can accept the .STL file directly, whereas others require *preprocessing*. Preprocesses include verifying the .STL file, slicing, and setting up parameters for machine control. Preprocessing is usually done at a computer separate from the RP system to save time and to avoid tying up valuable machine time.

After the .STL has been preprocessed and saved into a new slice file format, the new file can then be transferred to the RP system. File transfer can be done several ways, from manually transferring by disk or tape to network transfer. Since more complicated files are usually very large, a local area network or Internet connect is now almost essential for easy file transfer.

Once the final file formats are transferred to the RP device, the *build process* occurs. Most RP machines build parts within a few hours, but can run unattended for several days for large parts.

Upon completion of the build process, *post processing* of the part must occur. This includes removal of the part from the machine, as well as any necessary support removal and sanding or finishing.

If the finished part meets the necessary requirements, the cycle is complete. Otherwise, iterations can be implemented in the CAD file and the cycle is repeated.

1.5 Where the Technology is Today

As you can probably see, RP is a very powerful utility for prototype modeling in today's high-speed design-to-market workplace. Quick and inexpensive prototypes are the reason the technology was invented, as it is still a very important function to accommodate. RP is an excellent prototype-run alternative, but since its creation, as all good technologies do, it has branched out into even more useful applications, including casting, tooling, reverse engineering, and direct hardware fabrication.

1.5.1 Direct Use

With the stronger plastics and even metallic materials used in some of the RP processes, parts can be produced that will withstand respectful amounts of stress and higher temperature ranges. These parts can be applied in applications such a transonic wind-tunnel testing, as will be described in a later case study; snap-fit components, such as clasps or buckles; electronics devices; medical devices; and so on. The capability to print a part in hours and plug it directly into an application is a powerful advantage to any manufacturing, design, or similar business. Figure 1.2 shows a laser-welding shield built with RP, which was directly used to shield the laser cartridge from the bright flashes caused by welding.

Figure 1.2 This 3-inch laser shield was prototyped in approximately 2.5 hours.

1.6 A Sample Application of Rapid Prototyping

As a quick scenario, lets say a mechanical engineer has designed a pump housing for a certain type of racing engine. To get a prototype model conventionally, he would have to create numerous pages of detailed two-dimensional drawings, with various cross-sectional views, dimensions, and other specifications. He then would possibly create a three-dimensional model on the computer to allow for visual verification by the model maker.

The model maker would take the printed drawings and specifications and then proceed to cut the prototype housing out of a block of material using some type of machining process. After many man-hours of labor as well as a lot of time for transportation of the part from station to station within the shop, the model is returned to the engineer.

But alas! He discovers that the housing will not fit properly on the pump due to an unforeseen dimension of the surrounding parts. Now he has to start the process all over again. You can see where this can be very costly and time consuming.

With RP, the engineer could have created the three-dimensional CAD drawing (without any of the detailed two-dimensional drawings), downloaded directly to the RP machine in his office or lab, and then returned the next morning to find a prototype waiting for him to use.

Upon discovering any flaws in the design he can quickly update the computer model and download again to get the correct part built in a matter of hours, saving time, money, and a lot of heartaches. This enhances the process of concurrent engineering, or "getting it right the first time."

The obvious advantages of RP can thus be said to be speed, cost, and ease of mind. Many businesses are discovering these unique and powerful processes on a daily basis, including such professions as automotive, aerospace, medical, electronics, military, architectural, and more. And with such a wide cross section of techniques available, the processes can be molded and improved to fit each industry's needs.

1.7 Rapid Prototyping Processes

As the RP market expands, there are now many national and international companies manufacturing and selling RP processes. New technologies are being introduced at a growing rate, and thus this text will try to do them justice.

Each process has its own section, with a breakdown of the system operation, part finishing, typical uses, etc. Some systems have very short chapters, either due to the simplicity of operation or perhaps vendor-proprietary information. Also included are some case studies involving a cross section of the technologies used in wind-tunnel testing, investment casting, and other applications.

1. The JP-System 5 (JP5), by Schroff Development, builds models from CAD data using label paper and a knife plotter. JP5 is a simple and inexpensive modeler for creating rough three-dimensional models.
2. Ballistic Particle Manufacturing (BPM), now a historic component of the RP legacy, printed wax models by firing micro droplets of molten wax from a moving jet onto a stationary platform. BPM is currently not available due to a collapse of the manufacturer in late 1997, but is still held under a valid patent.
3. The Model Maker (MM), Model Maker II (MM2) and Rapid Tool Maker (RTM) by Sanders Prototype, Inc. produce highly accurate wax patterns using ink-jet printing technology with molten wax.
4. Multi-Jet Modeling (MJM), by 3D Systems, Corp., uses ink-jet printing technology with many jets enclosed into a single print head to produce concept models.
5. Direct Shell Production (DSP), used by Soligen, Inc., uses binder printing technology developed by MIT. The binder is printed onto layers of ceramic powder to produce investment shells directly from CAD.
6. The Z402 system by Z-Corp also uses MIT three-dimensional printing technology to build very fast concept models from a starch-like material. Also, the ProMetal system by Extrude Hone technology builds metal parts this way.

7. Fused Deposition Modeling (FDM), by Stratasys, Inc., produces models from wax or ABS plastic using motion control and extrusion technology similar to a hot glue gun. Also, the Genisys system uses FDM-like technology to build nylon concept models.

8. Laminated Object Manufacturing, by Helisys, Inc., builds physical models by stacking sheets of paper or plastic material and cutting away excess material with a laser.

9. Stereolithography, by 3D Systems, Corp., is the oldest RP system, and builds models by curing epoxy resins with a low power laser.

10. Selective Laser Sintering, by DTM, Corp., can build with a variety of materials, and works by selective melting together powder with a laser into a desire shape.

11. Laser Engineered Net Shaping, by Optomec Design Co., builds parts directly from metal powders, by fusing the powder together with a laser beam.

12. Other functional RP systems covered are Precision Optical Manufacturing, Laser Additive Manufacturing Process, and Topographic Shell Fabrication.

13. Brief coverage is given to international systems from Israel, Japan, Europe, and China.

1.8 Key Terms

Post processing. Cleaning operations required to finish a part after removing it from the RP machine.

Preprocessing. CAD model slicing and setup algorithms applied for various rapid prototyping systems.

Prototype. An initial model fabricated to prove out a concept or idea.

Rapid prototyping. Refers to the layer-by-layer fabrication of three-dimensional physical models directly from a CAD.

Solid modeling. The branch of CAD that produces true three-dimensional objects in electronic format.

Wireframe. An approximate CAD representation of a three-dimensional object.

Unit I

Concept Modelers

Concept modelers, often called office modelers, are a class of rapid prototyping (RP) system designed specifically to make models quickly and inexpensively, without a great deal of effort. The systems are usually small, inexpensive, quiet, and require very little or no training to operate. For these reasons, the systems are targeted to reside in design office environments, where they can ideally be operated much like a standard printer, only the prints from these systems are in three dimensions.

2

The JP-System 5

The *JP-System 5* (JP-5) is perhaps the simplest of all rapid prototyping (RP) processes. The JP-5 is nothing more (and nothing less) that a pen plotter that uses a knife blade and label paper instead of pens and draft paper. The JP-5 is manufactured by Schroff Development, Inc., and provides possibly one of the most economical bases for rapid prototyping on the market today. The system is a table-top unit, as is shown in Figure 2.1.

Figure 2.1. The JP-System 5 is a table-top unit that runs from a PC.

2.1 JP- 5 Hardware

The JP-5 operates from a personal computer as small as a 486/33 with 8 MB RAM (not included), and is very simple to use. The system comes in two sizes, the larger of which builds parts typically within a 12" x 18" cross sectional area. The JP-5 plotter is controlled through a serial port on the computer, by the JP-5 software.

2.1.1 Software

The software used to operate the JP-5 is actually embedded into a computer-aided design (CAD) package called Silver Screen. The JP-5 software accepts the standard .STL file, as well as Silver Screen CAD files. JP-5 slices the file and can be configured to run different materials as well as size ranges.

2.1.2 Build Materials

The build material supplied with the JP-5 is label paper. The label paper has adhesive on its underside, which is protected before use by a glossy sheet of backing. The paper comes in sheets that are 24" x 36" for the larger system. The JP-5 can use other materials that are in the same "peel-and-stick" configuration.

2.1.3 The Plotter System

The JP-5 plotter is a Graphtec single-pen plotter. The plotter is set up to either be sheet fed, as is the case with JP-5, or roll fed for standard pen plotting. The JP-5 is supplied with stainless steel cutting blades that fit in a modified pen holder. The pen holder allows the blade to be adjusted for cutting depth, and also gives the blade freedom to swivel. This swivel action, or free rotation, allows the blade to cut curvatures without a stepping motion. The plotter moves the paper back-and-forth in the −y direction, while the blade is controlled in the −x direction. This results in the two axis motion used to cut the cross sections of the parts being built.

2.1.4 The Assembly Board

The assembly board is a piece of three-quarter inch thick parti-cle board, and is used to align the sheets of paper for final fabrica-tion of the parts. It has many threaded holes across its surface, into which the alignment pins can be placed. There are several align-ment pins, which can also be attached end to end to make taller parts. Alignment pin configurations can be set in the JP-5 software, and the typical setting is for two or three alignment holes per slice. Also with the assembly board is an application board, which is a small board used to press the sheets together to get good bonding between layers.

2.2 JP-5 Operation/Build Technique

2.2.1 CAD File Preparation

The JP-5 starts with a CAD file, which can either be drawn up in Silver Screen or any other CAD package that puts out the stan-dard .STL file format. The JP-5 operating software is started up from Silver Screen, and the file you wish to build is opened. The CAD file must be converted from the three-dimensional format into horizontal two-dimensional slices, which will control the plotter motions. The CAD file is represented graphically in the software, and you have several operations you can execute before building the part. It is important to note that the style for assembling models on the JP-5 will vary from user to user, and the following is just one of those styles.

2.2.1.1 Scaling/Part Size

JP-5 allows you to scale parts to the size you wish to build. Parts can either be scaled uniformly across the $-x$, $-y$, and $-z$ coor-dinates, or be scaled in a combination thereof. When scaling down it is important to note smaller features that may not be distinguish-able at smaller sizes.

2.2.1.2 Orientation/Positioning

When building with the JP-5, the position of the parts can play an important role in the ease fabrication scheme. JP-5 allows you to rotate parts to a desired position before slicing and building. One consideration is the height of the part. If a part can be placed with the shortest dimension in the $-z$ direction, it will require the least amount of slices and therefore take less time and effort to put together.

Another aspect of orientation takes into account the shape of the part. It is usually effective to place larger cross sections near the bottom of the build, so that overhangs do not have to be supported. If a large cross section were to be placed on top of a significantly smaller one, the edges of the top slice would sag due to lack of physical support. This concept is demonstrated in Figure 2.2.

Curved surfaces provide another need for orientation. If it is practical concerning the number of slices, curves will be smoother if placed in the cutting plane. Sometimes a trade-off between curve accuracy and build layers can be achieved by rotating the part at a slant.

2.2.1.3 Slicing

Once the parameters are set, the software goes through a "slicing" function, which basically takes the drawing and cuts it into horizontal cross sections approximately 0.005" thick (the thickness of the paper). For larger parts, there may only be one slice per sheet of label paper. However, smaller parts have many slices per sheet. This will result is having sub sections of the part being built per sheet, which will then be assembled after all of the sheets have been cut. When the slicing is complete, the instructions can be downloaded to the plotter for building the part. Figure 2.3 shows the slices of a gear model in the JP-5 software.

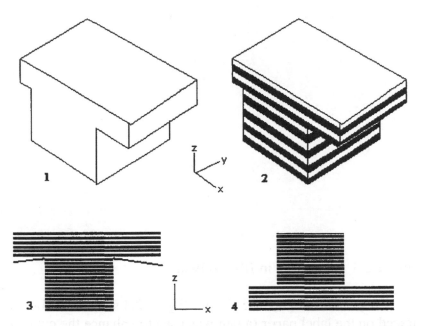

Figure 2.2   1) A sample of the need for supports when rapid prototyping a model. 2) A sample model with overhangs. 3) Exaggerated slice of the model in the given orientation. The ledges have nothing to keep them from falling. 4) Optimum orientation for the sample part.

2.2.2 Plotting

The plotter accepts the instructions from the computer one slice at a time, which allows the user to manually load paper for each slice. All slices except the mask are only cut partially through the label paper on the peel-away side, so that each cross section of the part can be peeled away from the backing and adhered to the previous layer. Alignment holes preset to match the peg scheme on the assembly board are cut completely through the paper and backing.

The first cut is a "mask" of the part, and is cut completely through the label paper and the backing. The mask is typically a rectangular stencil that encompasses the area of the initial slice of the actual part. The mask is used after the first slice is positioned and anchored.

Figure 2.3 Gear slices in JP-5 software.

Next, the first part slice is cut. Figure 2.4 shows the slice being traced on the label paper (a pen was used to enhance the cut edges). The software automatically mirrors the first slice so that it will align with the second and subsequent slices properly, since the remaining sheets will be placed opposite the first one. The backing is *not* peeled from the first slice. The first sheet is placed with the backing side down on the assembly pegboard with pins properly aligned through the alignment holes. This sheet is taped around the edges to the assembly board with masking tape to prevent any movement along the pegs. Figure 2.5 shows the first sheet being applied to the board.

The now upward-facing surface of sheet one has no adhesive for the next layer to stick to, so the mask is needed. The mask is placed on the assembly pegboard, using the registration pins, over the first slice, and spray adhesive is applied to the top of the first slice to give the second slice an adherent surface to start on. Figure 2.6 shows the mask being applied.

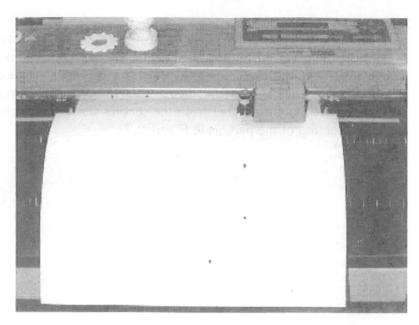

Figure 2.4 Gear slices being traced with the JP-5.

Figure 2.5   First slice of the gear placed on the assembly board.

The mask stencil is now removed, revealing rectangular patches of adhesive on the top of the first sheet. The part is now ready for the second layer. The second layer of paper is placed in the JP-5 and is cut like the first slice. After the cut, it is extremely helpful to peel the entire label from the second sheet *except* the actual cross section of the part, as in Figure 2.7.

Figure 2.6 Mask adhesive being applied to first gear slice.

Figure 2.7 Peeling away excess label around gear cross sections.

You now have the second sheet with only the part cross sections remaining, and you are ready to proceed to the next step. Since the label side is cut, but the backing isn't, placing the sheet label side down onto the square patch of adhesive will now allow you to peel the backing away, leaving the cross section attached to the first sheet, with its own adhesive side now exposed. Figure 2.8 shows the second layer being applied. Having the excess backing peeled from the second sheet keeps it from sticking to the spray adhesive in places undesired.

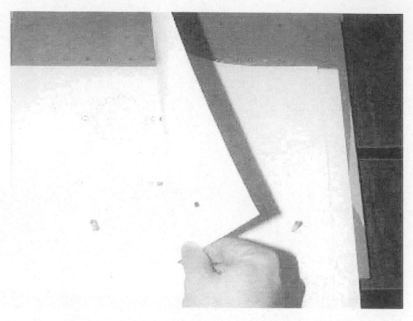

Figure 2.8 Applying subsequent layers to the gear model.

Now the third and all remaining slices are cut through the label side, the excess labels are peeled away, and the sheets are placed label side down and peeled to leave the cross section sticky side up for the next layer. The alignment pegs allow the cross sections to be set properly, ensuring a well-built part dimensionally. Keep in mind, however, that the part accuracy is only as good as the coordination of the operator.

Smaller parts may be built with several cross sections "per page." In this case the excess label removal and adhesion step may become a little more complicated, but it still works well. This process also requires subassembly steps, as it is only building sections of the part with the first sequence of sheet addition. The subassemblies typically have sub alignment holes so that they can be properly stacked together once the sheet peeling and adhesion step is completed. Figure 2.9 shows the finished gear (white), along with other parts made with the JP-5.

Figure 2.9 Finished JP-5 parts. All are approximately one-quarter inch thick.

2.3 Finishing a Part

Once the JP-5 part has been completely assembled, it should be sealed to prevent moisture from being absorbed into the paper. Moisture can cause the parts to warp, sag, or deform. Different sealers can be used, including sanding sealers and varnishes or polyurethane sprays, but the simplest method suggested by the manufacturer is to use ordinary school glue brushed on with a small paintbrush. The glue hardens, leaving a nice glossy seal around the part, which helps the appearance as well, however, care must be taken to protect the part from dirty surfaces before the glue dries. Figure 2.9 shows some finished JP-5 parts.

2.4 Typical Uses of the JP-5 Process

2.4.1 Preliminary Design

The JP-5 produces simple, inexpensive models that are best used in preliminary design environments, where dimensional accuracy and durability are less of an issue. It basically gives a designer a solid part he or she can hold and observe, as opposed to a flat drawing on a sheet of paper or computer screen. Unfortunately, the lack of automation tends to defer some designers and businesses, as the process would effectively consume at least one person's attention on a full-time basis, and most firms don't have the manpower to dedicate to spending so much time in producing simple models.

2.4.2 Education

Because of the attractive low pricing of the JP-5, it tends to often find its way into various educational facilities, typically on the collegiate level. It is an excellent demonstration of the concepts of layer-by-layer fabrication, because it allows the user the step-by-step interaction with the layers and cutting processes. The interface is easily taught and learned, and the student can get a feel (literally) of prototype modeling from CAD design. Students can design and fabricate mechanical components as class projects, for instance, which gives them experience in operating CAD systems for solid modeling, as well as designing for ease of manufacture. There are many universities now using the JP-5 as a teaching and learning instrument.

2.4.3 Research

On the research end, the simplicity of the system targets the JP-5 as a prime candidate for studying the effects of laminating stronger materials together in a rapid prototyping environment. Since the RP industry is in a constant shift toward the actual manufacturing of hardware, RP materials other than the ones currently used must continuously be studied, tried, and tested to meet the demands of the ever-growing needs of the manufacturing world. JP-5 allows one to experiment with basically any sheet-form material that can be cut with a knife blade. Results from these studies can lead to advancement in materials capabilities for the more automated RP processes, where lasers and continuous feed come into play, along with various other parameters.

2.5 Materials Properties

The JP-5 parts are actually fairly rugged once they are sealed with glue. Since the parts are only paper and glue, there haven't been many official tests done on the parts for tensile and compression strengths. The parts can survive being handled and dropped, but would probably not be used in many testing applications. For one, the dimensional tolerance relies heavily on the user, and at best probably is still approximately ± 0.1 inches. Also, any exposure to fluids or pressures may result in delamination of the paper between layers. Regardless, the application of verifying a concept visually is well filled by these parts.

2.6 Key Terms

Plotter system. For the JP-5, the plotter system consists of a computer and a modified Graphtec pen plotter with knife blades.

Assembly board. A large peg board with threaded holes at regular intervals that match pegs used for JP-5 part assembly.

Part orientation. The most crucial preprocessing step of the JP-5 system setup, consisting of placing the part in relative three-dimensional space whereas the build time and support geometry are both optimized.

For more information on JP-5 contact Schroff Development at 913-262-2664.

3

Ballistic Particle Manufacturing

BPM Technology, Inc., in Greenville, SC developed ballistic particle manufacturing (BPM). There are only a few BPM systems on the market due to the collapse of the company in late 1997, however the patent is still held by the inventor at Stereology Research, Inc. in Easley, SC. The systems that are used are mainly for rapid prototyping (RP) research, or for concept verification models. Ballistic particle manufacturing was marketed as one of the most affordable RP processes at that time, with very inexpensive build materials.

Ballistic particle manufacturing was represented as an "office modeler," since the build material was non toxic, the system was compact, and there was no need for solvents or post curing processes. Ballistic particle manufacturing's role in the RP environment was as one of the pioneers of ink-jet based, office-friendly systems, which several vendors now supply to the market.

3.1 System Hardware

The BPM Personal Modeler came with all hardware and software enclosed in a single compact unit. The footprint area of the entire system was only about 24" x 20", as it was expected to be used in an office environment. The BPM is controlled by a MS-DOS based 486/PC, which is housed within the unit on a unique slide and foldout tray. The system has a modem and a serial port for transferring files. The controlling software, which is used for slicing and building the models from the .STL files, is an MS-DOS™ based program that is set up for very little user interaction. Figure 3.1 shows the BPM system.

3.2 Ballistic Particle Manufacturing Operation

The BPM utilizes *ink-jet* or *droplet based manufacturing* (DBM) techniques, where it builds models by firing micro-droplets of molten wax material from a moving nozzle or jet onto a stationary platform. The platform is then lowered and the process is repeated until a three-dimensional object is printed. Most parts are built as a hollow shell, although down to a 0.25 inch internal cross hatch may be used.

Figure 3.1 The BPM Personal Modeler has all its components built into a slim cabinet.

3.2.1 Software

The BPM software is very easy to use for building parts. Essentially, the .STL file is opened, and if the file is in the proper orientation and scale, clicking the **Print** command starts the model building, just like printing to a regular desktop printer. After a prompt for a crosshatch size (or none), the BPM slices the file, homes out the machine, and begins printing.

3.2.2 Part Orientation

If the part does need manipulating before the build, the software has the capability to do so. Parts can be scaled, rotated, and/or translated to a desired orientation. The software will prompt if the part is too big for the build envelope, and will even give a scale factor, which will bring the part down to the maximum possible build size.

The part orientation in terms of support structures is less of an issue with the BPM. The unique five-plane build motion eliminates the need for much of the supports, and what little remain are automatically generated during the slice function. This build process is described later in the BPM Build Technique section.

Basically, the main concerns are surfaces with large radii of curvature, which should be placed in the −x, −y building plane for better curve definition. This orientation eliminates the stair-stepping effect of the layer-by-layer manufacturing process.

3.2.3 Crosshatching

The crosshatch setting represents the distance, or air gap, between the raster fill lines in a "solid" structure. For instance, if a solid cylinder .STL file is built with BPM, it will actually only build the walls and the ends, with an open egg-crate type structure internally. This egg-crate structure is the aforementioned raster fills, and thus the size of the openings is the crosshatch setting.

If the cylinder file were to be built with no crosshatch size specified, the BPM will only build the walls and the ends, much like an unopened soup container.

3.2.4 Calibration

The BPM is calibrated through the software as well. If parts are not building smoothly or correctly, the calibration sequence can be initiated fairly easily. This sequence prints a series of line patterns in the –x and –y direction, distinguished by printed numbers at each group of three lines. The line groups are viewed with a magnifying glass to determine which set is equally spaced, and then the corresponding number is entered in to correct the jetting properties.

3.3 Ballistic Particle Manufacturing Build Technique

The BPM process has five planes of motion for printing the wax material. The –x and –y motion is provided by a carriage system that moves the single jetting head front to back and side to side while printing. This action results in a thin layer of wax representing the contour of a cross section of the part. The z axis is provided by a platform mounted on screw leads, which enable it to be lowered to allow for the next cross section to be built. These three planes of motion are present in most RP systems.

The last two degrees of motion are what set the BPM system apart from others. The build jet can flip to fire almost completely horizontally. This motion is what eliminates the need for supports. As in the hollow closed cylinder example, the top surface can be jetted without supports because the flip motion of the jet allows material to be deposited sideways, in a horizontal plane. This unique capability is complemented with the capability of the head to rotate in the –x, –y plane for horizontal printing in the –x or –y plane.

3.4 Finishing of Ballistic Particle Manufacturing Parts

BPM parts are not intended to be finished, as they must be handled with care. The parts can be painted with a water-based paint using an airbrush, but that is about the extent of finishing that can be performed without damaging the part.

Figure 3.2 The BPM head flips to build at an angle, which eliminates most supports (Courtesy of BPM, Inc.).

3.5 Typical Uses of Ballistic Particle Manufacturing Parts

BPM parts are mainly used for concept visualization. Due to
the weakness of the material, the parts aren't well equipped for use
as functional components. The visualization aspect is a valuable in-
sight however, as a three-dimensional model can provide the necessary comfort zone in designing to manufacture.

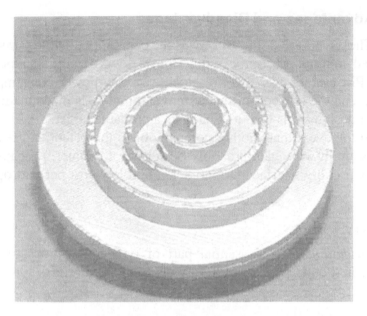

Figure 3.3 This compressor scroll model made by the BPM process was painted with water-based acrylic to give a metallic appearance.

Figure 3.4 These concept models were made using BPM process. The dime shows scale.

3.6 Advantages and Disadvantages

The advantage of the BPM was the capability to build parts that required minimal post-processing from a low toxicity standpoint. The BPM system was also presented as a low-power consumption device, and the low cost of the system and materials made the use in smaller design firms with low budgets a probable target.

In the event that another company buys the patent and revives the system, it will again probably be sold as a "printer," because of its user-friendly interface, non toxicity, and low cost per part operation.

3.7 Key Terms

Part orientation. An essential part of typical RP processes, the BPM could build parts in most any orientation with its 5-axis capability.

Cross-hatching. A time and material saving technique, in which parts are built with a hollow shell surrounding a corrugated grid structure. Now used in several RP techniques.

Calibration. In order to maintain repeatability and accuracy, ink-jet based RP systems, including the BPM, must go through a standard print calibration of a series of lines or intersecting lines that are then checked by the user and adjusted accordingly.

Finishing. Typically consists of sanding, polishing and painting. The BPM parts were too fragile to perform these operations, except for water-based airbrush painting.

4

The Model Maker Series

The Sanders Model Maker (MM) series captures the essence of the ink-jet printing technology, and builds in a layer-by-layer fashion, similar to other rapid prototyping (RP) systems. The MM uses several different types of data file formats but has only one base type for the build and support materials, wax. The MM was developed by Sanders Prototype, Inc. (SPI), a subsidiary of Sanders Design in Wilton, NH, in the early 1990s with the intention of revolutionizing the industry as it pertains to accuracy and precision.

4.1 Model Maker System Hardware

The MM system has evolved through three "models," Model Maker (original model), Model Maker II (MMII, second generation), and recently Rapid Tool Maker (RTM). The original modeler has a build envelope of 7" x 7" and the MMII has an envelope size of 13" x 7", whereas the RTM has a 12" x 12" working area. While both MMs are desktop models, the RTM is a self contained unit with an onboard computer. Figure 4.1 shows an MM unit and an MMII unit.

4.1.1 Software

Both modelers utilize MW (MW) software, manufactured by SPI, to prepare and manipulate the incoming file for use in the MM machine. The software can be operated through a variety of workstations, from UNIX to PC , and the current modeler has an onboard computer that can function alone after it receives the prepared file from a "dummy" PC whose sole purpose is for file slicing and preparation.

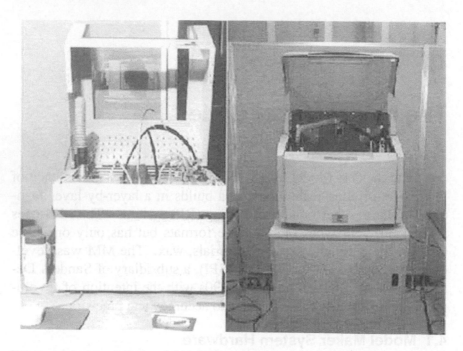

Figure 4.1 The MM and MMII by Sanders Prototype, Inc.

4.1.2 Build Materials

Both models use a build and support material to produce a 3-D model. These materials are wax based with the support having a lower melting point than the build. This insures that during post-processing, the support material will melt away leaving only the part, made of build material. Each material has its own heated reservoir and is very sensitive to contamination, which means handling should not occur (See Figure 4.2).

4.1.3 The Print-head

The print-head assembly consist of the print-head, print-head cap, purge spout, purge spout cap, cable, and saddle (see Figure 4.3).

There are two print-heads, one for building the part and the other for generating the necessary support. This support depends on the geometry of the part and can be produced around the entire part or just on certain areas. The jets sit on a carriage that enables them to move in the X and Y direction (left to right), while the stage moves in the Z direction (up and down). There are two processes that enable the materials to be transported to the print-heads. 1) Material is pumped to the feed lines by compressed air within the reservoir during the purge operation 2) There is an actual siphon that is conducted from the reservoir to the feed lines, to the print-head during the model build. The feed lines are heated, as are the print-heads. This heating of reservoirs, feed lines, and print-heads is necessary to have a continual flow of material.

Figure 4.2 Build and support reservoirs for the MM.

Figure 4.3 Print-head assembly of the MM.

4.1.3.1 Print-head Cap

The print-head cap is comprised of the top that closes the jet off and the connecting portion of the jet to the feed lines. This cap must be screwed on tightly to make sure the jet compartment is air tight and the connection to the feed line must be secure if there is to be a consistent and even flow of material.

4.1.3.2 Print-head Purge Spout and Cap

The purge spout is very important. It enables the user to purge the jet to assure there is a proper amount of material within the jet. If there is a air bubble in the material it is also necessary to purge by connecting a plastic tube to the spout and implementing the purge command for 2 seconds. Immediately after, remove the tube and replace the cap to eliminate excess air entering into the print-head body. This process should be done prior to every build and as the primary method to correcting a printing malfunction of the jet assembly.

4.1.3.3 O-ring

The o-ring is a seal between where the print-head cap screws into the print-head body assembly. It prevents leaking of material and the introduction of air into the body chamber.

4.1.3.4 Print-head Body

The print-head body is where the extra material is stored until it is printed onto the substrate surface. This is a very small space, therefore, an air bubble could cause a jet interruption and possibly a failure.

4.1.3.5 Saddle

The saddle secures the entire print-head assembly to the carriage. There is a locking device on the MMII that helps insure that the print-head is properly positioned to produce optimal printing results.

4.1.3.6 Tip

The tip is the orifice through which the material is printed. Never rub or touch this orifice with your hands or any other substance, this will damage the jet and cause it to be inoperable. The proper substance to use to remove any debris from the tip is the corner of a Kim wipe.

4.2 Model Maker Operation

4.2.1 CAD File Preparation

Prior to actually building the part, the STL file must be translated into the software language used by the MM. This software is MW and is used for the purpose of preparing and manipulating the file so that the MM can build it. The file, after being read into MW, produces a picture of the file on screen in the Cartesian coordinate system (−x , −y , and −z). A box appears around this grid with a bar that has many functions that allow the user to put the part in its desired orientation. From this box you can perform slicing functions, zoom functions, layer thickness alterations, part positioning, part sizing and other build parameters. The MW software is very useful and gives the user ultimate control over the end product. Figure 4.4 shows an STL file as viewed by MW.

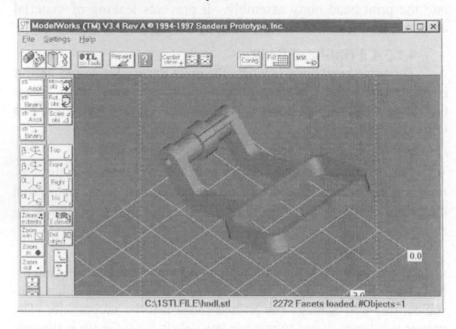

Figure 4.4 The STL file is viewed with MW before preparation to run.

4.2.1.1 Positioning the Model

There are several factors to consider when positioning the part. Among them are the distance the cutter travels, special features, detailing, opening edges, time to build, and the quality of the model. All of these can be changed as it pertains to the specific characteristics desired for the part. The fundamental rule for positioning is to have the longest length of the part parallel to the cutter.

4.2.1.2 Configuration Selection

This is a very important factor prior to building a part. This particular feature is accessed through the **Config** button on the MW window. This notebook contains the database of settings that determine how the model is built. Different parameters can be set with five tabs on the screen. These tabs are: Configuration, Units, Machine, Memory, and Build.

4.2.1.2.1 Configurations

Under this tab you are able to scroll through several configurations for slice thickness for each machine type (MM or MMII). Each configuration has an ID number that allows you to customize the slice thickness depending on the model being prepared. Figure 4.5 shows the configuration grid.

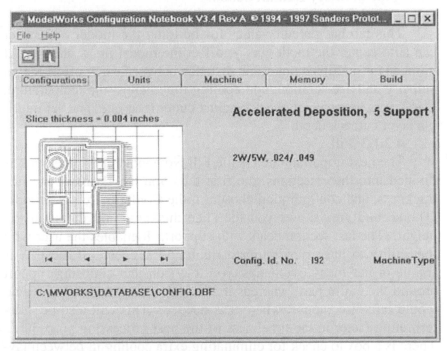

Figure 4.5 The configuration notebook in MW.

4.2.1.2.2 Units

The Units tab allows you to select the units in the configuration program and alters the way dimensions are interpreted in the geometry files. The choices are inches, millimeters, and centimeters.

4.2.1.2.3 Machine

The Machine tab allows you to select the current machine you are building on (MM or MMII).

4.2.1.2.4 Memory

This tab allows the user to select the amount of memory depending on the part size and complexity. The larger and more intricate the model, the more memory required. The machine requirements are at least 16 MB of RAM and preferably 32 MB of RAM. To make it simple most parts can be sufficiently completed by selecting the **Memory Default** button.

4.2.1.2.5 Build

This tab has default values for building the model. The user can auto center the model in –y, offset the model in –x and –y; set number for multiple copies and distance between each; offset the jet velocity and acceleration; add cooling time; adjust the temperature; skip layers between jet checks; adjust cutter feed rate; and set multiple layer cutback depth.

4.2.1.3 Fill

The next important area is the Fill Notebook. This notebook is divided into three sections, the first tells you the ID number, slice thickness, and configuration chosen in the Configuration Notebook. The second area shows you the slice thickness and the start/stop height. The last section deals with support. It enables the user with the ability to have maximum to minimum support. If the part is fragile or has many overhangs, you may consider choosing support around the entire part, whereas if it is a more rigid/sturdy part you would only use support where it is needed. This will also be a determining factor in the timeliness of the post-processing procedure. There is a box to check for eliminating extra cooling in between layers, the number of initial support layers, and continuing extra support to the top of the model. In addition, you can also reinforce the support walls by adjusting the excess-support control button.

4.2.1.4 Slicing

The MM utilizes Bview as a viewer of the .MM2 and .BIN files generated by MW. This viewer displays slice cross sections and their respective fill patterns. A digital readout at the bottom of the screen gives the ability to extract measurements from this file. This information is helpful in determining the files integrity and getting a view of slice-by-slice formation. This slice-by-slice file is the code the machine will use to generate a layer-by-layer creation of the model. To access this function you must select the **Bview** button from the menu. There are several different functions that you can implement to render the model and file exactly how you desire it to be prior to building. The Navigation buttons allow you to view the model slice by slice or at 10 percent increments, the Automated Control button gives you a real time build slice by slice, the Zoom buttons allow you to adjust the view of the model on the screen, and the Pan buttons allows you to adjust the –x and –y plane views of the model. Together, all of these functions give the modeler complete control over not only how the machine will build the part, but the customization of the part prior to the build.

4.2.1.5 Send to Model Maker

After slicing and orientation, the model must be sent to the on-board computer of the MMII. In order to do that the operator must choose the MM⟹ button, which will send the complete file to the MMII as Job.MMII. In order to send the file you must choose a file name, the printer, and then select the **OK** button. Once this is done it is now time to physically prepare the machine to build the model.

4.2.2 Building a Part

Once the part has been delivered to the MMII it is time to prepare the machine for building. Initially, you can check the material reservoirs to determine if you need to add any build or support materials. You can get a graphical representation by selecting <1> on your opening screen. The computer will tell you if additional material is needed and how much to add. Once you have added the material(s), allow 45 minutes for the material to be liquefied in the reservoir before use. But while you are waiting you can check the optical tape receptacle to make sure it is empty and you can mill the substrate. To do so, select <3> from the initial screen (Run MM), select <I>, and then <N>, this will allow you to choose the mill command and level your substrate. Mill the original surface (dull finish) of the substrate until it has a clean, bright finish. This ensures that the surface is level. The next and most important step is to check the jet-firing status. Before each use, perform a manual purge to refill the jet reservoir with material and make sure that the proper amount of air is within the reservoir also. Cut a 3-inch piece of plastic tubing, remove the purge cap, and place the tube on the purge spout. Hold the cylindrical tube over the tube and under <M> choose the respective jet you are purging (build or support). Once the jet has been selected, another menu will appear that will prompt your actions, from this menu, choose the **purge** command. Allow the jet to purge until you get an even flow of material into the tube container and allow it to flow for 2 to 3 seconds, then press any key to stop the purge. Immediately remove the tube from the spout and reapply the cap. After making sure that the jets are firing properly, go back into new build, select the file you want, and build.

4.2.3 Post-processing

The post-processing procedure is a process that must be monitored very carefully. When setting your initial temperatures you must be careful because the support material has a lower melting point than the build material. Either you can use a porcelain bowl-like container, a hot plate, and a thermometer, or you can purchase a sonicator with heat control and a built-in digital thermometer. If you purchase the latter, remember that the sonication produces its own heat, so additional heat may or may not be necessary depending on the part size. Post-processing is a hands-on process that involves time and attention. Allow the part to sit in the recommended VSO solvent solution at 35° C for 30-minute increments. Depending on part size you may want to play with the temperature settings and the time you allow it to soak in VSO. You want the support material to be mushy so that you can easily remove it with a tool of your preference (be careful not to destroy part surface). When all the support (red) material has been removed, you may re-finish your surface, paint it, or leave it as is. Remember, this process takes time, if you rush it you could sacrifice the integrity of your part.

4.3 Advantages and Disadvantages of the Model Maker

The power of the MM family of systems lies primarily with the production of small, intricately detailed wax patterns. The jewelry and medical industries have capitalized on this advantage due to their needs for highly accurate, small parts.

Perhaps the most apparent drawback of these systems are the slow build speed when it comes to fabricating parts larger than a 3-inch working cube. This is primarily due to the fact that the layer thickness required to achieve the high-dimensional accuracy is typically an order of magnitude thinner than standard RP systems. The result: the parts take 10 times longer to build.

Many concerns have been raised over the progression of the technology as to system operation reliability issues. More often than not, however, jet failures and failed part builds can often be traced back to a misunderstanding of the process operation and an overloading of the machine's true capacity.

4.4 Key Terms

Print head. The print head is the heart and soul of the MM systems, as most other components of the hardware can be obtained off-the-shelf.

Configuration. The main venue in the MW software for changing operating parameters of the MM system.

Fill. A large contributor to the final build time and material required for a part, the fill can be varied from solid to hatched.

Post-processing. The final cleanup of models after removal from the MM machine, post-processing requires several steps and direct user intervention in order to acquire a finished project.

For more information on the Model Maker, please contact Sanders Prototyping at 1-603-429-9700.

5

Multi Jet Modeling

The ThermoJet and the Actua 2100, both made by 3D Systems in Valencia, CA, fall into the growing area of the rapid prototyping (RP) market known as concept modeling. Both systems apply the Multi Jet Modeling (MJM) build style to produce wax prototypes with an array of ink jets. The systems are one of the least expensive in the line of RP technologies. They are also safe and clean enough to operate directly in a design office environment. The ThermoJet is actually a replacement to the Actua 2100, but there are still several Actua units in operation throughout the world.

5.1 System Hardware

The ThermoJet/Actua 2100 comes as a single self-contained unit (excluding the control computer), and is about the size of a large photocopy machine. One of the more unique features of the ThermoJet/Actua 2100 is the ability to network the system much like a standard printer. This allows the download of build files into a queue from various areas of the office. Parts are "printed" out in the order they are received the same way that documents are printed out on the standard printer. The ThermoJet/Actua 2100 also does not require any post-processing units, as the support removal is done easily by hand.

The key component of the MJM process is the material delivery system. The wax billet is loaded into a reservoir inside the cabinet of the machine. The material is kept molten there and is siphon-fed to the multi-jet head. The Actua 2100 multi-jet head has four rows of 24 jets each, a total of 96 jets for print-on-demand capability, whereas the ThermoJet systems have over 300 jets spanning the entire cross section of the part build area for a faster build capability.

5.2 Multi Jet Modeling Process Operation

5.2.1 System Software

The software for the ThermoJet/Actua 2100 system is very user-friendly, where user input is kept to a minimum. It is available on the PC platform, and accepts the standard STL RP file format. Basically all of the slicing and operating parameters are default settings that normally do not have to be changed. Parts can be nested in the –x, –y plane with a unique auto-nesting capability, and are then essentially "printed" into functional, accurate wax parts. Figure 5.1 shows the latest Thermojet system.

Figure 5.1 The ThermoJet MJM System (Courtesy of 3D Systems).

5.2.2 Build Technique

The MJM process builds parts by printing thin consecutive layers of the molten wax in the shape of the part cross sections. Like most RP systems, the parts are built onto a movable z stage, which lowers as the part is "printed." Currently, the Actua 2100 system prints with a layer thickness of 0.0039 inches, or three passes of 0.0013 inches, whereas the ThermoJet system prints multiple passes of thinner layers for higher resolution. The multi-jet head traverses in the x axis direction (left to right), as the printer gantry that houses the head increments the width of the affective print area in the y direction. The multiple jets are turned on and off where needed at precise intervals, which allows for an accurate final article in the xy plane. The final z dimensions suffer a slight amount of inaccuracy on the lower side due to support removal, yet the upward facing surfaces have excellent surface quality. Figure 5.2 shows the MJM process.

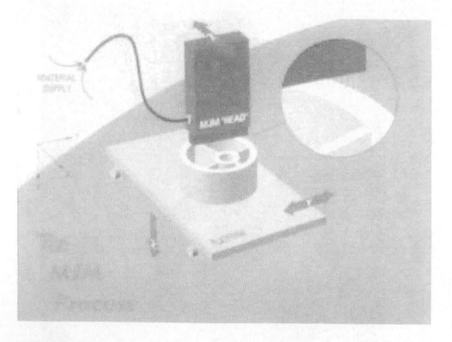

Figure 5.2 The multi jet modeling process (MJM) (Courtesy of 3D Systems).

5.2.3 Post Processing

MJM process parts are fairly easy to post process. The supports are built with the part build material, yet are strategically shaped for easy removal. The entire support structure consists of very fine columns of the build material that reach from the build platen up to any overhanging surfaces. These columns are thin enough that most of them can be "rubbed" off the part with your finger, and the rest can be sanded away. Again the upward-facing surfaces already have excellent surface quality (detailed enough to print a full-scale business card with raised print), so all of the post processing needed is on the underlying surfaces where the supports have been removed. For best results, the parts can be placed into a freezer for an hour prior to support removal. This causes the fine supports to become brittle and hence much easier to rub off. Care must be taken not to fracture the part itself, however.

After the quick support removal, the parts are complete. They require no post curing, infiltrating, or dipping. They are nontoxic and can be handled immediately after processing.

5.3 Typical Uses of Multi Jet Modeling

The MJM system was designed specifically for concept modeling in the office environment. The models can be built quickly and effectively, and are durable enough to demonstrate designs in presentations and meetings. The MJM process parts are also dimensionally stable enough to use for limited fit-check analysis applications.

Another use of the MJM process parts developed later for a more practical test hardware application. The wax makeup of the build material makes it applicable as an investment casting pattern material. The glossy surface finish and easy melt-out provide for clean, crisp metal castings.

As materials develop for the MJM process systems, they may begin to play a larger part in the more functional prototyping roles. Until then they will continue to serve an important component in the concept-modeling and casting realm. Figure 5.3 shows some sample MJM parts.

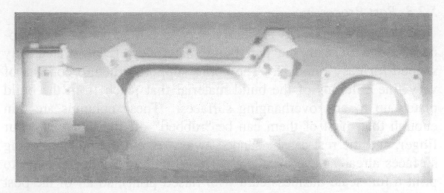

Figure 5.3 Sample parts from the MJM process (Courtesy of 3D Systems).

5.4 Advantages and Disadvantages of Multi Jet Modeling

The advantages of the MJM systems include speed, affordability, reliability, ease-of-use, cleanliness, crisp resolution, and networking capability. All of these features stack up to make for a very robust RP system.

The key disadvantages are not unlike those experienced by other RP systems, which include low-strength material, rough surfaces on the support material side of parts, and the moderately high cost of the build material.

Unfortunately, also, even though the post processing required has been significantly reduced in the Thermojet system, sometimes designers still find the effort required to clean up the parts is more than what fits into their working schedules. Thus the systems may tend to be used less than what would be ideal to maximize the design-to-manufacturing experience in a corporation.

5.5 Key Terms

Ink-jet technology. The same basic technology as used in desktop publishing printers, ink-jet technology provides the core process for concept modeling RP systems.

Networking. In this case, networking refers to electronic file transfer networks, such as local area networks (LANs), over which STL data can be transferred to the RP system. The MJM systems maximize the use of networking by allowing print queues, in which operators from remote computer workstations can send multiple parts to the systems to be manufacturing consecutively.

Concept modeling. Applying fast, inexpensive RP systems to the concurrent engineering design loop by providing physical verification models to designers, salesmen, etc.

For more information on MJM systems, contact 3D Systems at 805-295-5600.

6

3D Printing (Z402 System)

Three-dimensional printing, or 3DP, is an MIT-licensed process, whereby liquid binder is jetted onto a powder media using ink jets to "print" a physical part from computer aided design (CAD) data. Z Corporation (Z Corp) incorporates the 3DP process into the Z402 system. The relatively inexpensive Z402 is directed toward building concept-verification models primarily, as the dimensional accuracy and surface roughness of the parts are less than higher end systems. The initial powder used was starch based and the binder was water based, however now the most commonly used powder is a new gypsum based material with a new binder system as well. Models are built up from bottom to top with layers of the starch powder and binder printed in the shape of the cross sections of the part. The resulting porous model is then infiltrated with wax or another hardener to give the part dexterity. The Z402 is the fastest modeler on the market, with speeds 5 to 10 times faster than other current rapid prototyping (RP) systems.

6.1 Z402 System Hardware

The Z402 is currently available in only one size, which can build models up to 8" x 11" x 8". The overall size of the modeler is approximately 3' x 4', so it can fit in a fairly confined area. Parts built with the starch material can be hardened to fit the application necessary. Wax infiltration gives the parts some strength but also leaves them usable as investment casting patterns. Stronger infiltrants, such as cyanoacrylate, can be used to provide a durable part that can survive significant handling.

Since the starting point of this writing, Z Corp has advanced their 3DP system in several ways. First, they released updated print cartridges (Type 3) that last longer along with stronger infiltrants for durable parts. Secondly, a new material and binder system called ZP100 Microstone was released that provides stronger models directly from the machine with little or no postprocessing or infiltrant. Finally, an automated waxer was released that helps control the wax infiltration process if necessary.

The modeler has several important components, including the following:

1. <u>Build and Feed Pistons.</u> These pistons provide the build area and supply material for constructing parts. The build piston lowers as part layers are printed, while the feed piston raises to provide a layer-by-layer supply of new material. This provides the z motion of the part build.

2. <u>Printer Gantry.</u> The printer gantry provides the xy motion of the part building process. It houses the print head, the printer cleaning station, and the wiper/roller for powder landscaping.

3. <u>Powder Overflow System.</u> The powder overflow system is an opening opposite the feed piston where excess powder scraped across the build piston is collected. The excess powder is pulled down into a disposable vacuum bag both by gravity and an onboard vacuum system.

4. <u>Binder Feed/Take-up System.</u> The liquid binder is fed from the container to the printer head by siphon technique, and excess pulled through the printer cleaning station is drained into a separate container. Sensors near the containers warn when the binder is low or the take-up is too full.

The Z402 is operated through the COM port of a PC Workstation (not included), although the system has an onboard computer that can be used for diagnostics if necessary. The Z Corp slicing software is provided with the purchase of a Z402 system, and is compatible with Windows 98 and Windows NT.

Z Corp also sells a postprocessing package necessary for detail finishing and strengthening of the parts produced by 3DP. The package includes a glove box with air compressor and air brushes for excess powder removal, a heating oven to raise the temperature of the parts above that of the wax infiltrant and a wax-dipping unit that melts the wax and provides a dipping area for the parts. Figure 6.1 shows the Z402 system.

Figure 6.1 The Z-Corp Z402 3D printing system.

6.2 Z402 Operation

The Z402 has a very user-friendly interface, where very few commands are necessary to build a part. Since the parts are built in a powder bed, no support structures are necessary for overhanging surfaces, unlike most other RP systems.

6.2.1 Software

The Z402 starts with the standard STL file format, which is imported into the Z Corp software where it is automatically sliced and can be saved as a BLD (build) file. When a file is first imported into the software, it is automatically placed in an orientation with the shortest −z height. This is done as the fastest build capability, like other RP systems, is in the −x, −y direction. The part can be manually reoriented if necessary for best-part appearance. Multiple STL files can be imported to build various parts at the same time for maximum efficiency.

The default slice thickness is 0.008", however the value can be varied to fit the needs for particular parts.

Objects can be copied, scaled, rotated or moved for optimum part build. Moving/translating a part can either be done by a simple drag-and-drop method, or else by entering coordinates. Parts can also be justified to either side of the build envelope, be it front, back, left, right, top, or bottom, with a simple menu command. Parts are copied simply by highlighting the part and clicking one copy command. The new part is automatically placed beside the current part if there is room in the build envelope, otherwise it is placed above it.

Since the build envelope is a powder bed, three-dimensional nesting can be accomplished so that parts can be built in floating space to make room for others. This 3D nesting capability is only available in a few other RP systems, and provides for a higher throughput of parts to be accomplished.

After the STL is imported and placed, a "3D Print" command is issued and the part file is sent to the machine to build. During the build, a progress bar shows the percentage of the part building, as well as the starting time and the estimated completion time. When a build is complete, a dialog box is displayed with the final build time of the part, along with the volume of material used and the average droplet size of the binder used. The Z Corp software is shown in Figure 6.2.

6.2.2 Machine Preparation for a Build

Before the part can be printed, the machine must be checked and ready. The feed piston should have sufficient powder added, and the build area is landscaped by the wiper blade until it is level with powder. The binder fill and take-up must be checked, although the one-gallon containers typically last several months. The vacuum bag, which collects the overflow powder, is typically the most frequently changed item, necessary every few days. Excess powder/dust around the printer gantry and throughout the chamber is vacuumed away for prolonged operation. Also, for optimum performance of the print jets, a very small amount of distilled water is squirted into the jet cleaning station (aka the "car wash") on the printer gantry.

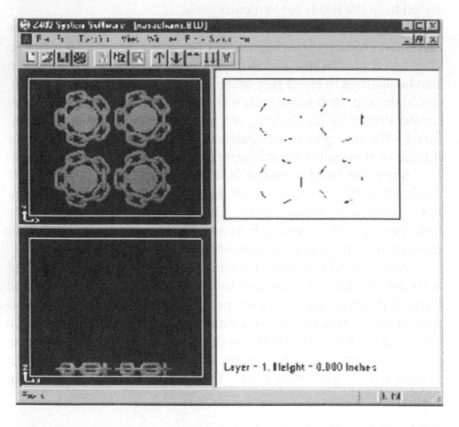

Figure 6.2 The Z Corp software allows for multiple-part placement, as well as operating the Z402 during a build.

6.3 Build Technique

The Z402 builds parts in layer-by-layer fashion, like other RP systems. The following set of figures (Figures 6.3 through 6.9) shows the sequence of part-building steps in the Z402 system, with details in the captions.

Figure 6.3 First, blank layers of powder are spread as a starting point for building upon. This step is controlled manually by the operator during machine setup, and is referred to in this text as landscaping. After this step, the machine is brought online and the remaining steps are performed automatically.

Figure 6.4 Next, the bottom cross section of the part is printed.

Figure 6.5 The feed piston is raised to supply more powder.

Figure 6.6 The printer gantry spreads the next layer of powder.

Figure 6.7 The next layer of the part is printed.

Figure 6.8 Subsequent layers are printed one after another.

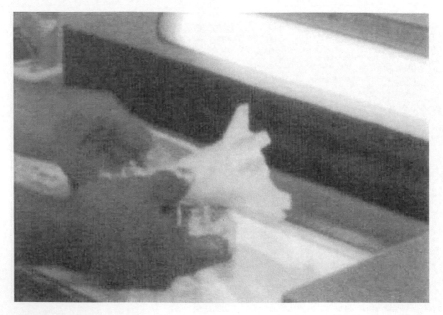

Figure 6.9 The final part is removed from the powder, ready to be postprocessed.

6.4 Postprocessing

Other than the Z402 system itself, there are several components needed for postprocessing of the part. For a concept model, the starch parts are generally infiltrated with paraffin wax, although more durable materials are available, from plastics to cyanoacrylate. Before infiltration, starch parts are fragile and must be handled with care. The following are the postprocessing steps for a part to be infiltrated with wax, with a total process time of about 15 to 20 minutes.

1. Powder Removal. After the parts are taken from the machine, the excess powder must be removed. With the system comes a small glove box with an airbrush system inside. The airbrush is used to easily and gently blow the powder off the part, and a vacuum cleaner is hooked to the glove box to remove the powder as it is blown from the part. (*5 Minutes*)

2. Heat for Infiltration. Once the powder is removed from the part surfaces, the part is placed in a small oven and heated to a temperature just above that of the infiltrant wax, to provide a wicking characteristic as opposed to coating. The part temperature for paraffin infiltrant is approximately 200°F. (*10 Minutes*)

3. Infiltration. Immediately after the part is heated, it is dipped for a few seconds into a vat of molten wax, then removed and placed on a sheet to dry. After drying the part is complete. (*5 Minutes*)

The actual postprocessing time will depend on the complexity of the part, the skill of the user, and the infiltrant used. Nonetheless, it is still minimal compared to some other RP processes.

The following set of figures (Figures 6.10 through 6.13) demonstrates the paraffin wax postprocessing technique generally used on Z402 parts.

Figure 6.10 Excess powder is removed with the aid of an airbrush.

Figure 6.11 The part is then heated 10 minutes at 200°F.

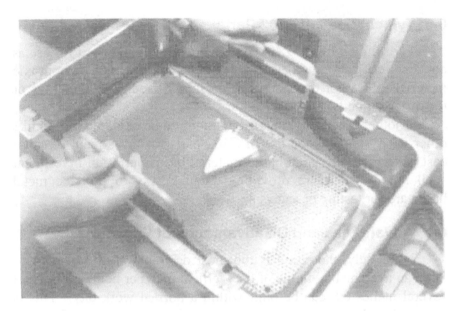

Figure 6.12 The part is dipped for a few seconds in the molten paraffin wax bath.

Figure 6.13 The dipped part is allowed to cool and dry.

6.5 Typical Uses of Z402 Parts

Parts built with the Z402 system are directly intended for use as concept-verification models in a design environment. The nontoxic materials allow for the models to be safely handled in meetings or the office, directly after fabrication.

Another application that is beginning to be explored, not unlike other RP systems, is the use of Z402 parts for investment or sand-cast patterns. The starch-based material burns out of an investment shell readily, therefore providing a quick way to produce metal hardware for testing or analysis.

6.6 Advantages and Disadvantages of the Z402

Ultimately, the speed is the most desirable trait of the Z402. With an average build time of one vertical inch per hour, even a part several inches tall can be built within a normal work day. This is extremely advantageous to any company where time is a factor in sales or production.

The key disadvantages of the system include rough part surfaces, which can be remedied with sanding, and the cleanliness problems faced when dealing with any system that uses a powder as a build material or operating medium. Also, the ink-jet cartridges must be replaced quite frequently, on the order of every 100 hours of operation, so users must understand that the jets are expendable items just as the build powder itself. Finally, these concept models aren't fabricated to high dimensional tolerances, which may hinder the building of complex assembly prototypes.

6.7 Key Terms

Ink-jet technology. The same basic technology as used in desktop publishing printers, ink-jet technology provides the core process for concept modeling RP systems.

Green strength. The handling strength of parts immediately after they are removed from the RP system, prior to any postprocessing.

Infiltration. A step required in some RP systems, where a matrix material is wicked into a porous RP pattern to enhance handling strength.

Powder removal. RP patterns fabricated in a powder bed often have excess powder that adheres to the surface of the part upon removal from the RP system. Different techniques are employed to remove the powder, in the case of 3D Printing an airbrush is used.

Self-supporting build. A unique feature available in powder bed RP technologies that allows parts with overhangs and internal structures to be built without the need for supporting trusses as are required in most RP systems.

Fastest RP technology. As of the last quarter of the year 2000, the Z402 systems maintain the fastest build times of any RP system that is commercially available.

For more information contact Z Corporation at 617-628-2732.

7
The Genisys Desktop Modeler

The Genisys (and Genisys Xs) system, produced by Stratasys, Inc. is an office-friendly modeling system that builds parts with a durable polyester material. The current line of Genisys systems are small, compact table-top rapid prototyping (RP) machines that deliver single-material capability, and interoffice network queues for operation much like a printer. Figure 7.1 shows the Genisys system.

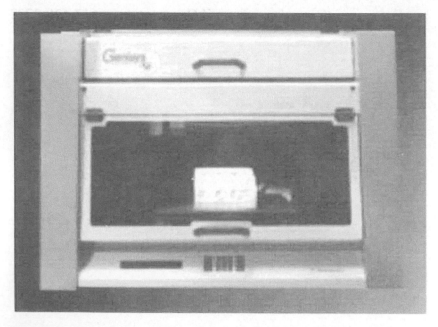

Figure 7.1 The Genisys Xs system produced by Stratasys, Inc.

7.1 History of the System

Not unlike most newly developed technologies, the original Genisys machines had small quirks and technicalities that prevented it from really being a true "trouble free" office modeler. However, after analyzing and working with customers, most of the systems were recalled and refurbished to correct the problems. The new line of Genisys, the Xs, apparently has printer-like reliability and operation, providing concept-modeling capability to the office environment as intended.

7.2 System Operation

7.2.1 Software

The software of the Genisys systems, which is compatible on both Unix and NT platforms, is designed for ease of operation. With simple point-and-click part-building features, the software automatically places, slices, generates supports, and then downloads the part file to the network queue to be fabricated. Parts can be set to be scaled automatically as well, although there is a manual scaling feature. Multiple parts may be nested in the −x, −y plane, again with single-click operability.

7.2.2 Build Material

The current build material is quoted as a "durable polyester". Since the systems have only one extrusion tip, the support structures are built of the same material, requiring mechanical removal upon completion of the part.

7.2.3 Hardware

The Genisys has a maximum build capacity of 12" x 8" x 8", whereas the entire system occupies a space of only 36" x 32" x 29". The unit weighs in at about 210 pounds and can operate on standard house current of 110 to 120 Volts AC.

The polyester material comes stock in the form of wafers, which are loaded into a bank of cartridges within the machine. One wafer is loaded into the deposition head, where it is melted and deposited in thin layers through a single extrusion tip while tracing the cross section of the part being built. Once the wafer in the head is spent, it is replaced by another automatically and the build resumes.

The build chamber is operated at ambient temperature, and fabricated parts can maintain dimensional accuracy in the range of ± 0.013 inches.

7.3 Typical Uses of Genisys Parts

The intended application of the Genisys system's product was mainly concept modeling and verification. However, as with all RP devices, various users have progressed the use of Genisys models into analysis, direct use, even low-impact wind-tunnel modeling. The material is said to be suitable for painting, drilling, and bonding to create the necessary appearance for an application.

7.4 Advantages and Disadvantages of Genisys

The advantages of the Genisys system include the ease of use and the network operability. Since the preprocessing is kept to a minimum, and the systems can be networked much like printers, the Genisys lends itself to the office modeling environment.

Perhaps the major disadvantage of the system would be its single-material capacity, which results in more difficult support removal on complex parts. This situation may well be addressed in the future, similar to what was done in the progression of its sister technology of fused deposition modeling, however the vendor has no plans released at the time of this writing.

For more information on Genisys, contact Stratasys, Inc. at 612 937-3000.

Unit II

Functional Modelers

Functional modelers are higher-end rapid prototyping (RP) systems that build parts larger, more accurately, and more durable than the office modeler systems. Functional modelers are usually larger and more expensive, and often times are more suitable for shop-floor or laboratory operation as opposed to a design or office environment.

8

Fused Deposition Modeling

Fused deposition modeling (FDM) is an extrusion-based rapid prototyping (RP) process, although it works on the same layer-by-layer principle as other RP systems. Fused Deposition Modeling relies on the standard STL data file for input, and is capable of using multiple build materials in a build/support relationship. FDM was developed by Stratasys, Inc. of Eden Prairie, MN, in the early 1990s as a concept modeling device that is now used more for creating casting masters and direct-use prototyping.

8.1 Fused Deposition Modeling System Hardware

The FDM systems have evolved through several models, beginning with the original 3D Modeler, a floor unit, and progressing through the various "desktop units", including the 1500, 1600, 1650, 2000, 8000, and Quantum. Basically, the 1500 through 2000 models are capable of building parts in the 10" x 10" x 10" range, whereas the 8000 and the Quantum can build 24" x 20" x 24" parts. Figure 8.1 shows an FDM 2000.

Since the beginning of this writing, Stratasys has released the FDM 3000 system, which has a unique Water Soluble Support (WSS) material. The WSS allows for the construction of more complex geometry and internal structures. Complicated support structures that would have previously been difficult to remove can now be flush away with a water-based solution. The FDM 3000 system also offers a larger build envelope than the 1500 through 2000 systems.

Figure 8.1 The Fused Deposition Modeler 2000 by Stratasys, Inc.

8.1.1 Software

All of the machines use the powerful QuickSlice (QS) software, manufactured by Stratasys and SDRC, to manipulate and prepare the incoming STL data for use in the FDM machines. The software can be operated on various types of workstations, from UNIX to PC based, and the modelers can either be operated directly from the workstation or by a "dummy" PC whose sole purpose is to free up time and space on the workstation.

8.1.2 Build Materials

The FDMs can be equipped to build with investment casting wax, acrylonitrile butadiene styrene (ABS) plastic, medical grade ABS thermoplastic, and/or Elastomer, although the ABS is currently used the most. The build and support materials come in filament form, about 0.070 inches in diameter and rolled up on spools. The spools mount on a spindle in the rear or side of the machine, and the filament feeds through a flexible tube attached to the back of the extrusion head. Figure 8.2 shows build material spools loaded on the FDM.

Figure 8.2 The build materials for FDM are stored on spools.

8.1.3 The Extrusion Head

The extrusion head is the key to FDM technology. The head is a compact, removable unit (good for materials changeover and maintenance), and consists of the following crucial components. Figure 8.3 is a schematic of the extrusion head that shows the various components described.

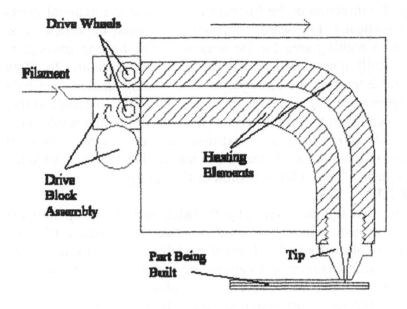

Figure 8.3 The key component of FDM technology is the extrusion head shown here.

8.1.3.1   Drive Blocks

The *drive blocks* are the raw-material feeding mechanisms, and are mounted on the back of the head. The drive blocks are computer controlled and are capable of precision loading and unloading of the filament. They consist of two parallel wheels attached to a small electric motor by gears. The wheels have a plastic or rubber tread, and are spaced approximately 0.070 inches apart and turn opposite to one another. When the wheels are turning and the end of the filament is placed between them, they continue to push or pull the material, depending on the direction of rotation. When loading, the filament is pushed horizontally into the head through a hole a little larger than the filament diameter, which is the entry to the heating chamber.

8.1.3.2   The Heating Chamber

The heating chamber is a 90-degree curved elbow wrapped in a heating element, which serves two primary functions. One is to change the direction of the filament flow so that the material is extruded vertically downward. Secondly, and most important, is to serve as a melting area for the material. The heating element is electronically controlled, and has feedback thermocouples to allow for a stable temperature throughout. The heating elements are held at a temperature just above the melting point of the material, so that the filament passing from the exit of the chamber is in a semimolten state. This allows for smooth extrusion as well as tight control on the material placement. At the end of the heating chamber, which is about 4 inches long, is the extrusion orifice, or tip.

8.1.3.3 Tips

The two tips are externally threaded and screw up into the heating chamber exit, and are used to reduce the extruded filament diameter to allow for better detailed modeling. The tips are heated by the heating chamber up to above the melting point of the material. The tips can be removed and replaced with different size openings, the two most common being the 0.012 and 0.025 inch sizes. The extruding surface of the tip is flat, serving as a hot shearing surface to maintain a smooth upper finish of the extruded material. The tip is the point at which the material is deposited onto a foam substrate to build the model.

8.1.4 Build Substrate

The foam substrate is an expendible work table onto which parts are built. The substrate is about one-inch thick and is fastened into a removable tray by one-quarter-inch pins. The pins are inserted horizontally through holes in either side of the tray, and pierce about two inches into the substrate to stabilize it during building. The substrates can sometimes be used several times for smaller parts by selectively placing them on unused sections, and by flipping them over to use the other side of the foam. The foam used is capable of withstanding higher temperature, as for the first few layers of the part the hot extrusion orifices are touching the substrate.

Modelers higher than the 1500 model have two drive blocks, heating chambers, and extrusion orifices in the head with independent temperature and extrusion control to accommodate two different materials. This allows for a build material, of which the part is made, and a support material. The support material is used to support overhangs, internal cavities, and thin sections during extrusion, as well as to provide a base to anchor the part to the substrate while building.

8.2 Fused Deposition Modeling Operation

8.2.1 Computer Aided Design File Preparation

Before building a part, the STL file has to be converted into the machine language understood by the FDM. The aforementioned QS software is used for this purpose. The STL file is read into QS, and is displayed graphically on screen in the Cartesian coordinate system (−x, −y, and −z). Also shown is the *bounding box*, a dashed three-dimensional box representing the maximum build envelope of the FDM. QS gives you options on the FDM system being used, the slice layer thickness, the build and support materials, as well as the tip sizes. Figure 8.4 shows an STL file as viewed by QS.

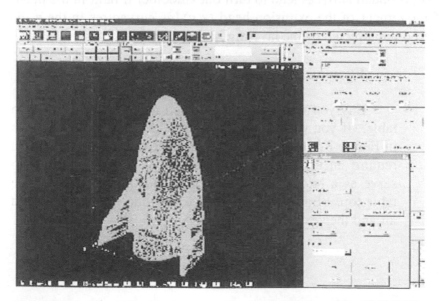

Figure 8.4 The STL file is viewed and manipulated by QuickSlice.

8.2.1.1   Part Size

First it must be affirmed that the part will fit into the bounding box; if not, it will either have to be scaled down to fit, or be sectioned so that the pieces can be built separately and then bonded together later. It is good practice for the designer to add alignment bosses and slots so that proper alignment of the subsections is achieved with ease. In some cases, for instance if the part fits in –x and –y but is too tall in the –z, QS can be used to section the part by slicing to a certain height, then starting a new build later at that height and finishing the part. This technique results in flat mating surfaces with no alignment bosses or slots, therefore it is up to the postprocessing person to align the subsections properly during bonding.

8.2.1.2   Orientation/Positioning

Once the part (or parts) has been deemed an appropriate build size, the part should be oriented in an optimum position for building. The shape of the part plays the major role in this, in that some orientations may require less supporting of overhangs than others. Also, rounded surfaces tend to turn out smoother if built in the plane of movement of the extrusion head (x,y), as curvatures in the z direction are affected by the layering build technique.

Example: Orientation of Table for Optimum Build.

Say you want to build a scaled model of a round patio table. If you were to build the table standing on its legs, as in Figure 8.5, the round top will come out nicely, but it will require an extremely large amount of support material to serve as a bridge for building the "floating" surface of the table top. (Basically, it is hard to extrude material into thin air and then expect it to stay there.)

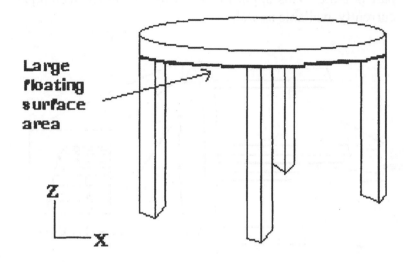

Large floating surface area

Z

X

Figure 8.5 The upright table requires excessive supports for the top.

Secondly, you could build the table on its side, as in Figure 8.6, which would require much less support material than before, in that it now only supports the thin floating sections of the legs. Now, however, the rounded table top will lose its definition due to the layering affect of the build process. You can see in the diagram that the layers create small "stair steps" as the curvature increases in the z plane.

The final, and best, option, is to build the table upside down (Figure 8.7). Again, the rounded top will have good definition due to the precision control of the head. Now the necessary support material has been minimized, as there are no floating surfaces that require support. Essentially, the only support material used will be for the anchoring base layers. This simple change of orientation has saved material, time, and accuracy! There is less support material required, which also cuts down on the build time, and the desired definition will be obtained. Of course, parts are usually never this simple, but nonetheless much time and cost can be spared if the build orientation is well thought out before the part is built. This applies to most all of the RP techniques currently available as well.

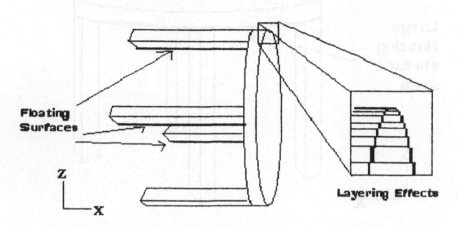

Figure 8.6 Orienting the table sideways reduces the necessary supports, however the definition of the circular area is lost due to the stair-stepping effect.

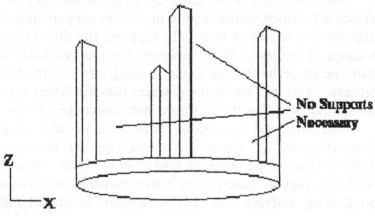

Figure 8.7 Upside down, the table requires minimal support material, gets finer definition on curvature, and also reduces the build time.

8.2.1.3 Slicing

Once the part(s) has been properly oriented and/or scaled, it must be sliced. *Slicing* is a software operation that creates thin, horizontal cross sections of the STL file that will later be used to create the control code for the machine. In QS, the slice thickness can be changed before slicing, the typical slices ranging from 0.005 inches to 0.015 inches. Thinner slices may be used for higher-definition models, but this increases the time required to complete a part build. Likewise, less accuracy-sensitive parts can be built much faster using a thicker slice value. There becomes a tradeoff between the desired accuracy and the time needed, with the optimum value determined by the user.

In QS, the slicing is shown graphically, and you can actually flip through the slices individually if you need to examine or edit them. Figure 8.8 shows a sliced aerospace vehicie from the front view in QS. (Resolution is low to protect software propriety).

Figure 8.8 QuickSlice slices the three-dimensional solid into horizontal cross sections.

QS allows you to perform simple editing functions on the slice files, so for example if you want to offset a set of slices to make a hole smaller by a given dimension, you can do so quickly without having to return to the original computer aided design (CAD) program. Also, the editing function allows the repair of minor flaws in the STL file, with the options of closing and merging of curves. Finally, a graphical representation allows you to inspect possible problem areas of the part for building. For instance, you may be able to detect surfaces that will be difficult to support in the current orientation. Once the slice file is in satisfactory condition, the file can be saved for future manipulation or reference.

8.2.1.4 Build Parameters

QS typically has optimum build parameters set as default for the slice thickness and material you chose, but it will also allow manual intervention so that you can vary several different settings. Some of the parameters later discussed can be tweaked to decrease build time, model weight, and the amount of material required for the build.

8.2.1.4.1 Sets

QS uses sets, or packages of build parameters. Sets contain all of the build instructions for a selected set of curves in a part. Sets allow a part to be built with several different settings. For example, one set may be used for the supporting structure of the part, one for the part base, another for the thicker sections of the part, and still another for exposed surfaces of the part. This allows the flexibility of building bulkier sections and internal fills quickly, while getting finer detail on the visible areas of a part. Sets also allow chosen sections of a part to built hollow, cross hatched, or solid, if so desired. Two of the build parameters commonly worked with are the road width and fill spacing.

8.2.1.4.2 Road Width

The road width is the width of the ribbon of molten material that is extruded from the tip. When the FDM builds a layer, it usually begins by outlining the cross section with a perimeter road, sometimes followed by one or more concentric contours inside of the perimeter. Next, it begins to fill the remaining internal area in a raster, or hatch, pattern until a complete solid layer is finished. Therefore, the three types of roads are the perimeter, contour, and rasters. Either of these can be turned on or off by QS. One good is example is for support structures. Typically, the perimeter and contours are turned off on supports, leaving a set of thin vanes that are easier to remove during postprocessing. Another example is to turn off the rasters, which allows you to build a hollow part, because the only material being extruded is for the walls of the part.

The road width can be as small as the diameter of the tip opening to approximately twice that size. The width is controlled by increasing or decreasing the extrusion rate in conjunction with the speed of the head, but dont worry, the software calculates all of that. Basically the user changes the width values to get a faster (if larger) or smoother (if smaller) part. Nominally, to maintain good surface finish on a part, the exposed surface road widths are best kept at a minimum, regardless of the interior road-width values.

8.2.1.4.3 Fill Spacing

Fill spacing is the distance left between the rasters or contours that make up the interior solids of the part. A fill spacing set at zero just means that the part will be built solid. But, QS allows the user to set values, therefore a bulky part can be significantly reduced in time and material cost by leaving air gaps between each consecutive road. Keep in mind that to make sure the part appears solid, all exposed surfaces need to be built with zero fill, or else you will be able to see the open spacing.

NOTE: Generally, an air gap of one to three times the width of the rasters works well, and about 10 layers of zero fill should be used when approaching exposed surfaces. This will effectively reduce the build time, while still maintaining a solid feel and appearance.

8.2.1.4.4 Creating and Outputting Roads

Once all parameters have been set, the roads are created graphically by QS. The user is then allowed to preview each slice, if so desired, to see if the part is going to build as required. Figure 8.9 shows a slice with roads as depicted by QS. The software actually shows the path of the extrusion tip for each individual road, therefore one can check for errors in the build sequence. For example, a wall may have been too thin for the chosen road width and QS left it blank. The operator must then go to the set that the wall is included in and lower the road-width values to make it fit. After satisfactory roads have been created, the data is written out as an SML file (Stratasys Machine Language), which is essentially a numerical control text file that can be read by the FDM.

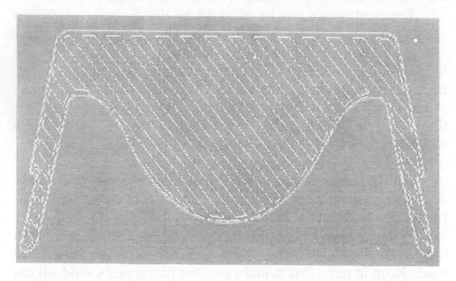

Figure 8.9 A close-up of a road slice in QS. Note the outline perimeter and raster fills.

8.2.1.4.5 Getting a Build Time Estimate

QS has a very good build-time estimator, which activates when an SML file is written. Basically, it displays in the command window the approximate amount of time and material to be used for the given part. A build estimate can also be acquired for previous SML files by opening them and simply clicking the Build Estimate button. Estimating the build times for parts is an important aspect in any industry that produces components on a schedule. This allows for efficient tracking and scheduling of the FDM system workloads.

8.2.2 Building a Part

The software setup requires most of the operator's time. Once the SML file has been created, it can be downloaded to the FDM through the parallel port of the PC just as if the file were being printed on an ordinary desktop printer. The FDM receives the file, and will begin by moving the head to the extreme −x and −y positions to "find" itself, and then raises the platen to a point to where the foam substrate is just below the heated tips. After checking the raw-material supply and the temperature settings, the user then manually places the head at the point where the part is to be built on the foam, and then presses a button to begin building. After that, the FDM will build the part completely, usually without any user intervention. Figures 8.10 through 8.15 shows a 6-inch hollow aerospace model being fabricated.

8.2.3 Finishing a Fused Deposition Modeling Part

FDM parts are usually some of the easiest rapid prototyped parts to finish. FDM features the Break Away Support System (BASS), which allows the support material to be peeled away easily by hand with a knife or pliers. The materials are easy to finish by sanding, and the ABS plastic parts may be made very smooth by wiping them down with a cloth moistened in acetone or a similar light solvent. The investment-casting wax parts may be smoothed with an electric hot knife or be dipped in a lower temperature wax such as paraffin. The parts tend to require very little finishing before they are ready to be delivered, depending on the application.

Figure 8.10 The FDM just before building an aerospace model. Notice the two extrusion tips are slightly buried into the foam substrate. This provides an anchor for the following part-build process.

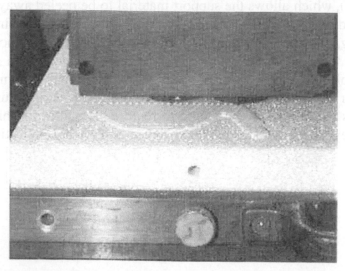

Figure 8.11 The aforementioned model after 15 minutes. The model is mostly support material due to a taper on the back of the design.

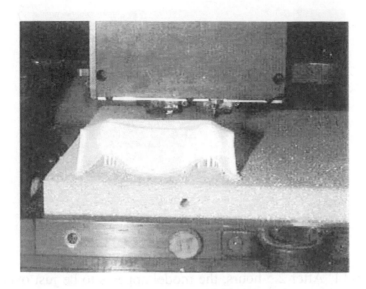

Figure 8.12 After 1 hour the model has built over 1-inch tall. Notice the part is being built hollow in order to speed up the process.

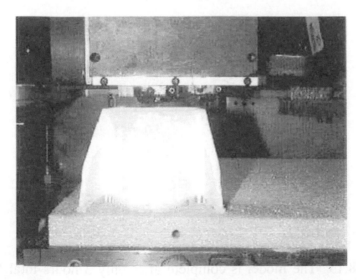

Figure 8.13 After 2 hours, the model is progressing steadily. The
time to build each layer will decrease from here to the finish because
the cross-sectional area continuously gets smaller on this part.

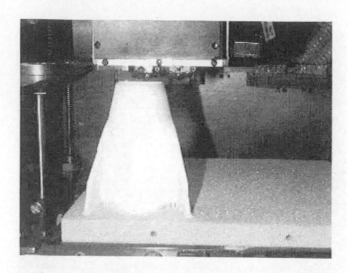

Figure 8.14 After 2.5 hours, the model appears to be just over half finished, although it lacks only about one half of an hour due to the dwindling cross-sectional area of the design.

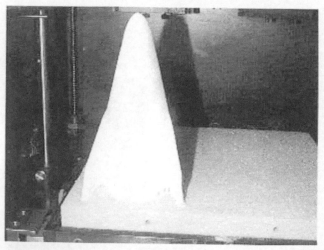

Figure 8.15 The model is complete after only 3 hours total. Machining the same model could easily take 10 to 20 man-hours, as well as shipping, tracking, and off-duty hours.

8.3 Typical Uses of Fused Deposition Modeling Parts

8.3.1 Concept/Design Visualization

Like other RP systems, the FDM systems provide an excellent route to obtaining prototype models for initial observation of a design. The ABS and Elastomer parts are rigid enough to survive handling and transporting from meeting to meeting, or even down to the shop floor. Parts can be made with various colors to represent different components of a system. The build materials are also relatively inexpensive, so models can be re-iterated more so than if prototypes were being machined or formed.

8.3.2 Direct-use Components

Due to the rigidity of the ABS parts, they can be used in various applications to replace traditionally machined, extruded, or injected plastic parts. The FDM can build directly usable electronics housings, low-speed wind-tunnel models, and working gear assemblies among other uses. This allows users to directly fabricate prototypes and test them before actually machining the final design.

8.3.3 Investment Casting

The investment-casting wax offered by FDM opens up yet another avenue of applications. If prototypes are needed in a metal form, the parts can be prototyped using the investment-casting wax, and then carried through the traditional investment-shell casting process to obtain usable metal components. The investment-shell process basically consists of shelling the wax part in ceramic, and then melting out the wax to have a mold into which molten metal can be cast. Hence, prototype castings can reduce design-to-market costs by getting it right before the final manufacturing step is initiated.

8.3.4 Medical Applications

The Medical Grade ABS has been approved by the U.S. Food and Drug Administration (FDA), and therefore is used by the medical industry to produce various parts within the industry. Since CAT

Scan and MRI data can be converted into the .STL file format, custom models of internal organs, bones, etc. can be reconstructed and studied before a patient ever goes into surgery!

8.3.5 Flexible Components

The recently released Elastomer material opens yet another dimension of functionality for the FDM systems. Flexible test components such as seals, shrouds, and tubing can be prototyped with the Elastomer material to proof out the concepts or "make the sale" on crucial designs.

8.4 Fused Deposition Modeling Materials Properties

As mentioned earlier, the FDM systems now have the capability to build parts with four different materials. Investment-casting wax (ICW06) is an industry-standard foundry wax that is used for many casting applications. ABS (P400) is a rigid plastic material that also comes in six colors: white, red, green, black, yellow, and blue. Medical Grade ABS (P500) has the strength of ABS but also can be sterilized to produce functional medical components. Elastomer (E20) provides a flexible build-material source that can be used for seals, gaskets, shoes, and other applications.

The materials properties were provided courtesy of Stratasys and are as follows:

Material	Tensile Strength, psi	Tensile Modulus, psi	Flexural Strength, psi	Flexural Modulus, psi
P400	5,000	360,000	9,500	380,000
P500	5,400	286,000	8,500	257,000
ICW06	509	40,000	619	40,000
E20	930	10,000	796	20,000

Figure 8.16 Mechanical properties of FDM build materials.

8.5 Advantages and Disadvantages

The strength and temperature capability of the build material is possibly the most sought-after advantage of FDM. Other major advantages include safe, laser-free operation and easy postprocessing with the new water-soluble support material.

Although significant speed advancements have been made with newer FDM systems, the mechanical process itself tends to be slower than laser-based systems, therefore lack of build speed is a key disadvantage.

Also, small features like a thin vertical column prove difficult to build with FDM, due to the fact that each layer must have a physical start-and-stop extrusion point. In other words, the physical contact with the extrusion tip can sometimes topple, or at least shift, thin vertical columns and walls.

8.6 Key Terms

Extrusion head. The key component of FDM technology, the extrusion head performs the material melting and deposition functions while being moved on the −x, −y carriage.

Drive blocks. Located inside the extrusion head, the drive blocks pull filament from the material supply spools and push them through the heated head chamber.

Raster fill spacing. Also known as the air gap, the raster fill spacing is the distance between each individual bead of material that is deposited. It can be set to zero to make a solid part, or opened up to build parts with internal cavities.

Slice thickness. The thickness value for each horizontal cross-section to be deposited, generally a value from 0.007 to 0.010 inches.

Water-soluble support. A late feature of the FDM systems, water-soluble support material is removed from the part by dissolving away, as opposed to the mechanical removal of standard support material.

Build substrate. A removable foam pad that is used to anchor parts steady during the FDM build process.

Road width. The width of each individual bead of filament material that is deposited, typically one to three times the width of the extrusion tip diameter.

For more information on FDM contact Stratasys, Inc. at 612-937-3000.

9

Laminated Object Manufacturing

Laminated Object Manufacturing (LOM), is a rapid prototyping (RP) technique that produces three-dimensional models with paper, plastic, or composites. Helisys, Corp. in Torrance, CA developed LOM, led by Michael Feygin. LOM is actually more of a hybrid between subtractive and additive processes, in that the models are built up with layers of material, which are cut individually by a laser in the shape of the cross section of the part. Hence, as layers are being added, the excess material not required for that cross section is being cut away. LOM is one of the fastest RP processes for parts with larger cross-sectional areas, which makes it ideal for producing larger parts. Figure 9.1 shows a LOM1015 system.

9.1 System Hardware

The LOM system is currently available in two sizes, the LOM1015 and the larger LOM2030. The LOM1015 can build parts up to 10" x 15" x 14", whereas the LOM2030 can build parts up to 20" x 30" x 24". Both operate using the same technique, and the most common build material today is paper. Parts built with LOM-Paper generally have a wood-like texture and appearance. The build material has pressure and heat-sensitive adhesive on the backing, and comes in various widths starting from 10 inches. Material thickness ranges from 0.0038 to 0.005 inches, about the thickness of two or three sheets of notebook paper.

The LOM operates from a PC workstation, which is provided with the LOM System when purchased. The LOMSlice™ software provides the interface between the operator and the system. LOM doesn't require a preslice of the STL file, that is, once the parameters are loaded into LOMSlice the STL file slices as the part builds. This process of continuous slicing is called *slice-on-the-fly*.

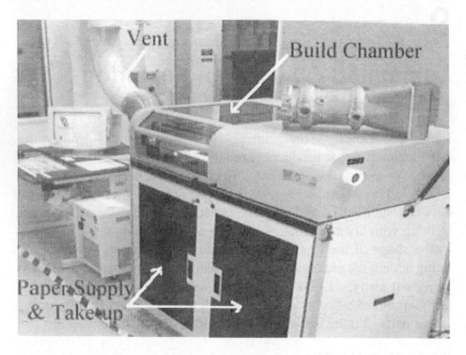

Figure 9.1 The Helisys LOM1015 machine.

The LOM has a feed spindle and a take-up spindle for the build material. The feed spindle holds the roll of virgin material, whereas the take-up spindle serves to store the excess material after a layer is cut. A heated roller traverses across the face of the part being built after each layer to activate the adhesive and bond the part layers together.

An invisible 25W carbon dioxide laser is housed on the back of the LOM, and is reflected off three mirrors before finally passing through a focusing lens on the carriage. The carriage moves in the −x direction, and the lens moves in the −y direction on the carriage, thus allowing the focal "cutting" point of the laser to be moved like a plotter pen while cutting through the build material in the shape desired. This −x and −y movement allows for two degrees of freedom, or essentially a two-dimensional sketch of the part cross section. The part being built is adhered to a removable metal plate, which holds the part stationary until it is completed. The plate is bolted to the platen with brackets, and moves in the −z direction by means of a large threaded shaft to allow the parts to be built up. This provides the third degree of freedom, wherein the LOM is able to build three-dimensional models.

Some smoke and other vapors are created since the LOM functions by essentially burning through sheets of material with a laser, therefore, the LOM must be ventilated either to outside air or through a large filtering device at rates around 500 cubic feet per minute (cfm). This is perhaps the most difficult part of installing a LOM1015 system, as the rest of the system runs on household current and requires no major facilities modifications. The larger LOM2030, however, has to be installed with plenty of operating space to allow for lift-truck loading of the large material rolls, as well as for recommended overhead-crane access for the removal of extremely large parts.

9.2 Laminated Object Manufacturing Operation

All of the previously mentioned hardware and software components work together to provide fast, economical models from the LOM system. The way the LOM constructs parts is by consecutively adhering layers of build material while cutting the cross-sectional area of the part with a laser. The LOMSlice software that comes with the LOM machine controls all this. The following description of operation is described with paper as the build material, but operation with other materials works in the same fashion.

9.2.1 Software

As with all RP systems, the LOM must begin with the standard RP computer file, or the STL file. The STL is loaded into LOM-Slice (Figure 9.2), which graphically represents the model on screen. Upon loading the STL file, LOMSlice creates initialization files in the background for controlling the LOM machine. Now there are several parameters the user must consider and enter before building the part.

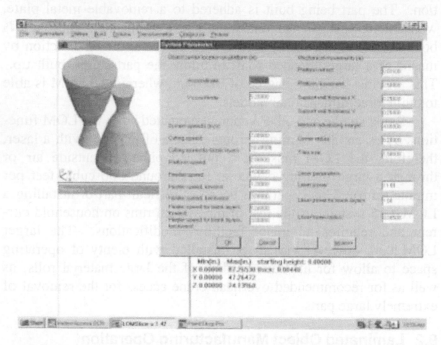

Figure 9.2 The LOMSlice software provides the interface between the operator and the LOM machine.

9.2.2 *Part Orientation*

The designed shape of the part or parts to be built in the LOM must be evaluated for determining the orientation in which to build the part. The first consideration for part orientation is the accuracy desired for curved surfaces. Parts with curved surfaces tend to have a better finish if the curvatures of the cross sections are cut in the −x, −y plane. This is true due to the fact that the controlled motion of the laser cutting in the −x, −y plane can hold better curve tolerances dimensionally than the layered effects of the −x, −z and −y, −z planes. If a part contains curvatures in more than one plane, one alternative is to build the part at an angle to the axes. The benefits here are twofold, as the part will not only have more accurate curvatures, but will also tend to have better laminar strength across the length of the part. Figure 9.3 shows the part orientation function.

A second consideration for part orientation is the time it will take to fabricate a part. The slowest aspect of the build process for the LOM is movement in the –z direction, or time between layers. This is mainly because after the laser cuts across the surface of the build material, the LOM must bring more paper across the top face of the part and then adhere it to the previous layer before the laser can begin cutting again. For this reason, a general rule of thumb for orienting long, narrow parts is to place the lengthiest sections in the –x, –y plane (lying down flat). This way the slowest part of the process, the actual laser cutting, is minimized to a smaller amount of layers.

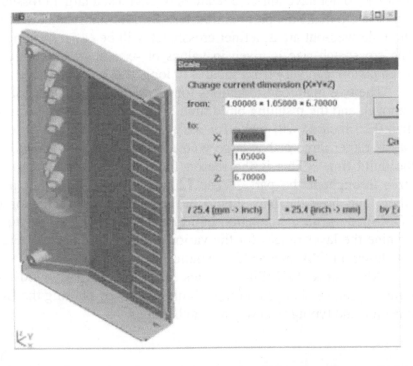

Figure 9.3 The orientation of the part must be thought out before beginning a part build on the LOM. LOMSlice allows the operator to scale, translate, or rotate the model for the optimum performance.

Parts can be scaled, rotated, or translated in LOMSlice to allow for the best arrangement as seen by the operator. There are some third-party software vendors that have automatic nesting functions that will strategically place parts in optimum orientations for the machine selected.

9.2.3 Crosshatching

As is described later in the build technique section, cross-hatching is necessary to get rid of excess paper on individual layers. Crosshatch sizes are set in LOMSlice by the operator and can vary throughout the part. Basically, the operator puts in a range of layers for which he wants a certain crosshatch pattern. For sections of the part that do not have intricate features or cavities, a larger crosshatch can be set to make the part build faster. But, for thin-walled sections and hollowed-out areas, a finer crosshatch will be easier to remove. The crosshatch size is given in values of $-x$ and $-y$, therefore the hatch pattern can vary from squares to long, thin rectangles.

The two main considerations for crosshatching are ease of part removal, and the resulting build time. A very small hatch size (less than 0.25 inches) will make for easy part removal, however if the part is rather large or has large void areas, it can really slow down the build time. This is the reason for having varying crosshatch sizes throughout the part. The LOM operator can either judge where and how the part should be crosshatched visually, or use LOMSlice to run a simulation build on the computer screen to determine the layer ranges for the various needed hatch sizes. Figure 9.4 shows a LOMSlice build simulation in progress.

Also, since LOMSlice creates slices as the part builds, parameters can be changed during a build simply by pausing the LOM machine and typing in new crosshatch values.

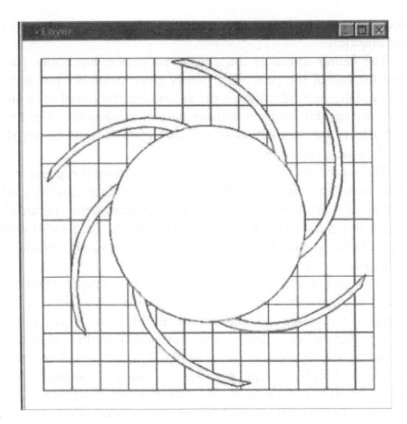

Figure 9.4 LOMSlice can be used to simulate a part build on screen to help verify crosshatch patterns.

9.2.4 System Parameters

There are various controlling parameters, some of which typically do not change from part to part, that are used each time the LOM is set up. The laser power, heater speed, material advance margin, support-wall thickness, and heater compression are some system parameters the operator has the ability to change if needed.

The *laser power* is a percentage of the total laser output wattage. For instance, the LOM1015 is usually operated at a laser power of about 9% of the maximum 25 watt laser, or approximately 2.25 watts. This value will be different for various materials or machines, but essentially it is set to cut through only one sheet of the build material.

The *heater speed* is the rate at which the hot roller passes across the top of the part. The rate is given in inches per second, and is usually around 6 inches per second for the initial pass and 3 inches per second for the returning pass of the heater. The heater speed affects the lamination of the sheets, so it must be set slow enough to get a good bond between layers.

The *material advance margin* is the distance the paper is advanced in addition to the length of the part. This is usually started out at about 1 inch to keep scorched paper from being included in the part, but can be changed to a lower value (~ 0.25 inch) during the part build to avoid excess buildup on the take-up spindle and wasted paper.

The *support-wall thickness* controls the outer support box walls throughout a part. It is not ideal to change this value during a build, although it is possible. The support-wall thickness is generally set to 0.25 inches in the –x and –y direction, although this value can be changed by the operator. For example, if a part is 0.1 inches too long for the build envelope, the user can make the support wall in that axis be only 0.15 inches to allow the build to take place.

The *compression* is used to set the pressure that the heater roller exerts on the layer. It is measured in inches, which is basically the distance the roller is lifted from its initial trek by the top surface of the part. Values for the compression will vary for different machines and materials, but are typically 0.015 to 0.045 inches.

9.3 Laminated Object Manufacturing Build Technique

Once all of the values have been plugged into LOMSlice, the LOM is now ready to begin building the part. Due to residues built up from burning material, the moving parts, lens, and mirrors must all be cleaned before beginning a part build.

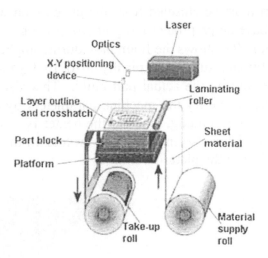

Figure 9.5 The LOM process (Courtesy of Helisys, Corp.).

Figure 9.5 shows a schematic of the LOM process. The paper roll is loaded onto the feed spindle in the bottom of the machine, and the paper is threaded around rollers, across the platen and then back down to the take-up spindle. An electrical device attached to a spring-loaded wheel, which is placed against one of the paper rollers, monitors the feed distance of the paper. This allows the system to feed only the necessary amount of paper to cover the cross-sectional area of the part or parts. A cleaned build plate is mounted on the platform, which is then raised to the top home position.

Next, a perimeter is cut out of the paper representing the largest cross-sectional area of the parts to be build. The blank is removed and a layer of double-sided foam tape is placed across the build plate, and cut with the laser to match the previous blank size.

The *support base* of tape not only serves as a thermal barrier between the plate and part, but also serves two other important functions. If the part were to be directly bonded to the plate, the removal thereafter would prove extremely difficult. The foam support layer, while still providing a steady anchor for the part during the build, also makes part removal from the plate easier. Secondly, since the part must be chiseled from the plate (even with the foam), the support base helps protect the part from being damaged during removal. Figure 9.6 shows the foam base during application.

For additional support, typically 5 to 10 layers of paper are added before starting the actual part build. This also gives the operator a chance to verify the machine parameters by checking how well the layers are bonding, whether the heater is scorching the paper, and so on. This also increases the protection to the part during final removal from the plate.

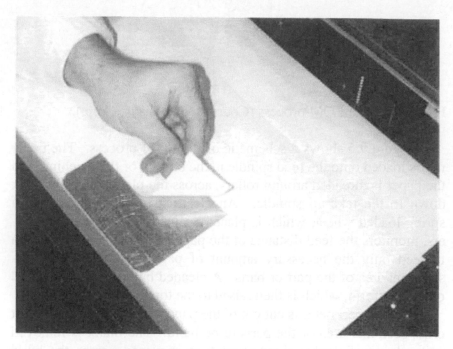

Figure 9.6 LOM parts need a foam base to anchor the part to the build plate during processing, as well as to ease in part removal.

After the support base is applied, the LOM starts building the part up from the bottom, as is designated in the software setup. The paper spindles turn to move new paper across the build plate. Next, the build plate is raised until it touches a sensor on the heater roller, where it stops to wait for adhesion. The heater roller now rolls along the surface of the paper, activating the adhesive backing while simultaneously applying downward pressure. The heater makes a return pass to its home position to finish the *adhesion step*, which is shown in Figure 9.7.

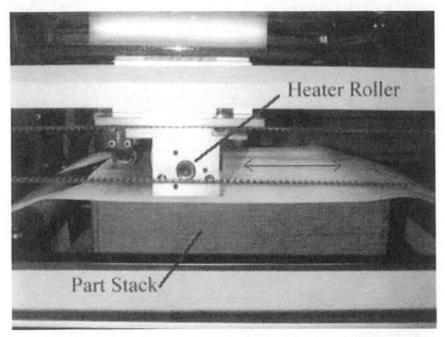

Figure 9.7 The adhesion step of the LOM process. A heated roller activates the backing of the paper to bond it to the previous layer.

Now the lens plots or traces the outline of the current cross section of the part, directing the laser to cut through only that sheet of paper in the desired shape. Each layer cut by the LOM must have a uniform, rectangular cross section, which is necessary to prevent the remaining waste paper from being adhered to the part. This means that if the actual cross-sectional area of the part being built varies, there will be some additional paper around the edges of the part necessary to fill out to the rectangular area. This waste paper can be reduced by "nesting", or strategically placing other parts in the STL file into such void areas to make good use of the excess paper. Otherwise, the excess paper on each layer is cut up into small crosshatches, to allow for easy removal of the part upon completion of the build. Also, for parts with internal cavities, the excess material inside is crosshatched as well. Figure 9.8 demonstrates the *cutting step* of the LOM process, in which the part cross-section, along with the crosshatches and support walls, is cut.

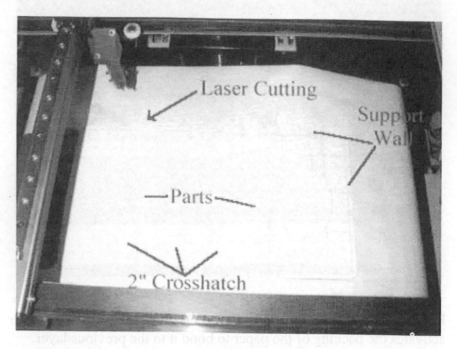

Figure 9.8 The LOM cuts through the paper layer with a carbon dioxide laser in the shape of the part cross section.

Finally, a rectangular perimeter is cut that encompasses the outer edge of the part and crosshatches, and a second perimeter is cut that is offset from the previous to the outside by a set number (generally 1/4 inch). These two outer perimeters form a support wall, which holds all of the crosshatched cubes and the parts within together throughout the remainder of the build.

After the final outer cut, the part is then lowered to allow the paper rolls to advance the remaining waste paper to the take-up spindle, thus bringing solid paper across the part surface again. In order to allow for the continuous feed of the paper, the paper is wider than the build area, therefore the paper between the feed and takeup spindle is never fully severed. Figure 9.9 shows the paper margin around the part area. The excess material going to the takeup spindle will consist of the remaining margins of the cut-out layers, which will be discarded after the build is complete. The adhesion step repeats, the laser cuts the next layer, and so on until the part is complete.

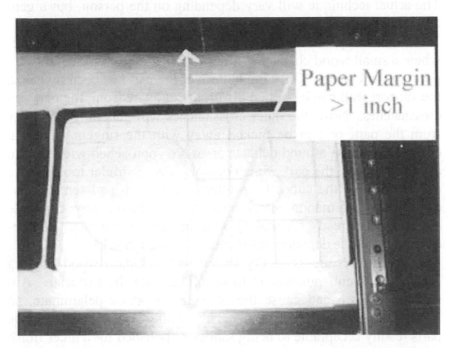

Figure 9.9 The LOMPaper is wider than the build area to allow for continuous feed capability.

The completed part when removed from the LOM machine is a rectangular block, because the outer support wall and crosshatched cubes are still positioned around the actual product. The part and plate are removed from the machine, where the part is then removed from the plate. A citrus-based solvent is sprayed on the plate to remove any remaining residue from the foam support layer, and the plate is then cleaned and ready for the next build. The cubes and the support wall are removed from the part by hand, which reveals the final product. This decubing process is described in more detail in the following section.

9.4 Finishing a Laminated Object Manufacturing Part

The support base, support walls, and crosshatches are removed from the finished part by a process often referred to as *decubing*. The actual technique will vary depending on the person, but a general technique follows.

First, the part block is placed upside down on a work table, where a small wood chisel or similar tool is used to remove the support tape and layers. Once the bottom of the part has been revealed, the support wall can be pulled off of, or cut away from, the part and crosshatches. Now, the outer crosshatches will generally fall away from the part, or can be picked away with the fingers. Internal hatches and those around delicate areas are approached with care, to avoid damage to the part. Small dental picks or similar tools can be used to remove the cubes from internal structures and sensitive areas. After some manual intervention, the decubing is now complete and the part is ready for sealing, sanding, and/or painting. The decubing process is depicted in Figures 9.10 through 9.14.

Paper parts are generally sealed with standard wood sanding sealer to prevent moisture from absorbing into the material. Absorbed moisture can cause the parts to distort or delaminate, so sealing is a necessary precaution. The sanding sealer also makes the parts readily acceptable to being sanded or polished for a nicer finish and to remove any remaining effects of the crosshatches. Finally, the parts can be painted after they are sealed, or else be left in the wood-finish form.

Figure 9.10 The part comes out of the LOM machine in a block.

Figure 9.11 The support-base layers must be removed with a wood chisel.

Figure 9.12 After base removal, the outer support wall is removed.

Figure 9.13 The loose outer crosshatches can be removed without tools.

Figure 9.14 The fully decubed LOM part can now be sealed and/or sanded for best finish.

9.5 Typical Uses of Laminated Object Manufacturing

The final LOM parts have a relatively good handling strength, except in very thin sections. They can be used for various applications, from concept verification to test prototypes. The LOMPaper parts can also be used as investment casting patterns, as well as masters for silicone-rubber injection tools.

9.5.1 Concept Verification

The attractive appearance of the LOM models, along with the capability for good surface quality, make the LOM models good candidates for concept-verification applications. Designers or engineers can have complete mechanical assemblies fabricated quickly on the LOM machine, and then use the models for design-review meetings, management briefings, and more importantly, to see if the model has all of the needed features and shapes as visualized with the computer design. If a design flaw is caught and prevented during this stage of the manufacturing process, it can provide millions of dollars in savings, as well as preventing months of unproductive development work.

9.5.2 Fit-check Analysis

As a follow up to concept verification, fit-check analysis is the next phase of RP. Assembly models can be put together to check for part interference, fit, and appearance. Also, models can be mounted on existing hardware to likewise check for appropriate fit and ergonomics. The inexpensive LOM models provide the visual link to checking out a design before advancing to the expensive manufacturing phase.

9.5.3 Direct Use Components

There are several ways in which LOM models can be used directly. Sometimes a model is all someone needs. One such example is for topography mapping from electronic surface data. Topographical maps can be made of planet surfaces from satellite data, or of underwater trenches and valleys and the like. These maps can be made accurately from digital data, as opposed to the conventional techniques of hand carving from sheet coordinates. This can allow unmanned underwater or interplanetary exploration to be recorded in a more useable fashion, and help to study and plan manned expeditions to such areas.

Another example of direct-use LOM models is in the medical industry. LOM models can be made from MRI or CAT scan data of bones, organs, and arteries. One example is for accident victims with skull or facial damage. These models can help doctors plan reconstructive surgery before they ever enter the operating room. Modeling scaled-up versions of viruses or tumors can aid in the study and treatment of such phenomena.

Finally, models can help serve as aids for machinists cutting the final designs. A three-dimensional replica is definitely easier to understand than complicated blueprints or even solid models on a computer screen. A physical model can serve as a powerful prevention tool against manufacturing errors.

9.5.4 Casting and Molding Patterns

For more durable hardware, LOM models can be a primary step to acquiring metal components. By using the models as patterns in a sand casting or investment casting process, metallic test hardware can be produced. If many parts are needed, the negative of the part needed can be built on the LOM, and then used as an injection tool to make many wax casting patterns from one LOM model. Yet another use is to pattern a silicone-rubber tool around the LOM model, then use this tool to inject wax or plastic parts. For limited-run production prototypes, the LOM models can provide a faster, more cost efficient way of testing out a new product design.

9.6 Advantages and Disadvantages

As seen throughout this chapter, the LOM advantage comes from the ability to produce larger-scaled models using a very inexpensive paper material. The finishing ability of the parts and the good handling strength couple with the speed and accuracy to provide an all-around quality modeling system. The materials are environmentally compliant and have not shown any capability of being health threatening.

Some possible disadvantages include the need for decubing, which is somewhat labor intensive for an "automated" process. Also, the emission of smoke and fumes, although vented out, can be a slight nuisance to visitors or tourists who aren't accustomed to it. And the fact that the machine operates by burning through paper can raise some concern to fire-safety officials.

Otherwise, the LOM is an all-around hearty RP system, and the prospect of more advanced materials on the horizon makes the LOM that much more desirable to have in a model shop or plant.

9.7 Laminated Object Manufacturing Materials Properties

LOM models can usually hold dimensional tolerances of about ± 0.010 inches and the materials properties as relayed by the vendor are shown in the chart in Figure 9.15.

Material	Tensile Strength	Elastic Mod	Elongation	Hardness
LPS 038 Paper	9500 psi	971 kpsi	2%	55-70 ShoreD
LPH 042 Paper	3710 psi	366 kpsi	10.70%	55-70 ShoreD
LXP 050 Plastic	12400 psi	500 kpsi	9.60%	n/a

Figure 9.15 Physical properties of LOM materials.

9.8 Key Terms

Laminated Object Manufacturing (LOM). RP process that build three-dimensional physical models from sheets of laminated material cut with a laser.

Slice-on-the-fly. The process of continuous slicing used by the LOMSlice software in which parameters can be changed during a part build.

Crosshatching. The process of cutting excess material around a LOM part into smaller pieces, which allows for the easier removal of the finished part.

Laser power. A percentage of the total laser output wattage, set to allow only one sheet of build material to be cut per layer.

Heater speed. The rate at which the hot roller passes across the top of the part, which affects the lamination capability of the sheets.

Compression. System parameter used to set the pressure that the heater roller exerts on the layer during the adhesion step.

Support base. The sequence of initial layers on top of a foam tape substrate, which is used to protect the part during removal as well as providing a thermal barrier between the part and the aluminum build platform.

Adhesion step. The time between layers when the heater is rolled across the top of a new sheet to glue it to the previous layer.

Nesting. The strategic placement of parts in three-dimensional space to allow for the smallest amount of waste in a LOM part build. Nesting hence increases the efficiency of the part-building process.

Support wall. A thin-solid shell cut around the outside perimeter of the LOM part and crosshatches to maintain stability in a part build.

Decubing. The manual removal of the support base, support walls, and crosshatches from a finished LOM part.

For more information on LOM, contact Helisys, Corp. at 800-470-7077.

10

Stereolithography

Stereolithography, as was noted in the introduction chapter of this book, was the first rapid prototyping (RP) process to reach the market in 1987. Produced by 3D Systems, Corp., in Valencia, CA, the Stereolithography Apparatus (SLA) has progressed through a long succession of models and advancements since its inception. While other vendors have begun to develop and market stereolithography systems, 3D Systems still maintains the majority of SLA machines in the field.

10.1 The Stereolithography Apparatus

Table 10.1 shows some of the various models and their build capabilities as of the third quarter of 2000.

Model	250/30	250/50	3500	5000	7000
Laser Type	HeCd	HeCd	Solid State	Solid State	Solid State
Laser Power	12 mW	24 mW	160 mW	216 mW	800 mW
Laser Life	2,000 hrs	2,000 hrs	5,000 hrs	5,000 hrs	5,000 hrs
Recoat	Blade	Zephyr	Zephyr	Zephyr	Zephyr
Min. Slice	0.006"	0.004"	0.002"	0.002"	0.001"
Beam Diam.	0.008"	0.008"	0.008"	0.008"	variable
Scan Speed	30 in/sec	30 in/sec	100 in/sec	200 in/sec	variable
Max Part Vol.	10x10x10"	10x10x10"	14x14x16"	20x20x23"	20x20x23"
Max Part Wt.	20 lb	20 lb	125 lb	150 lb	150 lb

Table 10.1 Approximate operating values of 3D Systems' stereolithography machines (Courtesy of 3D Systems).

10.1.1 Software

The SLA control and setup software has gone through various changes since the inception of the original MS-DOS version, which still operates on many SLA250 and SLA500 machines today. For a while, there were three packages required: a UNIX-based system for viewing and positioning (SLA View™); another UNIX-based third-party software for generating support structures (Bridge-Works™); and finally the slicing and system operation software (SLA Slice™) located on the PC attached to the SLA machine.

The next generation of software combined the view/positioning and the support generation into a more powerful UNIX-based software deemed Maestro™, but still maintained the same DOS software for operation of the system. The latest systems have Microsoft® Windows NT® software for all operations: 3D Lightyear™ for viewing and positioning, support generation, and slicing; and Buildstation™ for operating the SLA machine. Fortunately, the newer software can still write code for the old DOS-operated machines as well.

10.1.2 Build Materials

The SLA is a liquid-based RP process, which builds parts directly from CAD by selectively curing, or hardening, a photosensitive resin with a relatively low-powered laser. *Polymerization* is the process of curing a plastic or polymer by introducing a catalyst. In other words, polymerization links small molecules (monomers) to create larger chain molecules (polymers), this finally develops into a fully cross-linked solid polymer. *Photopolymerization* is essentially the same effect, only that the catalyst introduced is light energy. The light energy kicks off a free-radical polymerization, where the liquid photopolymer is phased from liquid to gel to solid. The solid obtained is, however a thermo-set, so it can only be used one time after it has been cured (nonrecyclable). In the SLA process, the light energy is introduced by a focused laser, which selectively cures the resin in a desired shape following a CAD file.

The original SLA build materials were acrylate based. They were improved upon by epoxy-based materials, also known as the

ACES (Acrylic Clear Epoxy System) build style. The epoxy materials provide advantages over the acrylate resins in that they have better materials properties and are less hazardous. The integration of the epoxies did require, though, longer exposure time for cure as well as higher-powered lasers.

There are now a wide variety of resins available not only from the vendor but also from third-party vendors as well. The competitive market continues to open up with higher-performance build materials at slightly lower costs. Resins can be purchased to improve resolution, temperature capacity, or even speed of the build.

10.1.3 The SLA Hardware

The build chamber of the SLA contains a removable vat that holds the build resin, a detachable, perforated build platen on a –z axis elevator frame, and an automated resin-level checking apparatus. The vat has a small amount of –z movement capability, which allows the computer to maintain the exact height per layer.

A recoater blade rides along a track at the top of the vat, and serves to "smooth" the liquid across the part surface to prevent a rounding of edges due to cohesion effects. Some systems now have a *Zephyr* recoater blade, which actually siphons up resin and delivers it evenly across the part surface.

In an enclosed area above and behind the build chamber, resides the laser and optics required to cure the resin. The laser unit is long and rectangular, about 4 feet long, and remains stationary. The laser beam is transferred to the part surface below by a series of optics, the final of which moves to scan the cross section of the part being built.

Also required, however, are the postprocessing units; an ultraviolet oven call the Post Curing Apparatus (PCA); and an alcohol bath large enough to hold entire build platens with parts attached. Parts are washed in the alcohol or a similar solvent immediately after being removed from the machine (while still attached to the build platen). This step removes any extra resin that clings to the surfaces of the part. After the final supports are removed, with some build styles the parts are required to be placed in the PCA to finish fully curing.

10.2 Stereolithography Apparatus Operation

The process begins with a solid model in various CAD formats. The solid model must consist of enclosed volumes, or be "water-tight," before it is translated from the virgin CAD format into the standard .STL file. The .STL is oriented into the positive octant of the Cartesian coordinate system, and then is translated up the −z axis by at least 0.25 inches to allow for the building of supports. These supports serve to attach the part to the build platen, and yet still allow for the safe removal of the part. The solid model is also oriented for optimum build, which involves placing complex curvatures in the −x, −y plane where possible, and rotating for the least −z height as well as to where the least amount of supports are required. More likely than not, only one or two of these recommended situations can be met, therefore the optimum trade-off must be made by the operator.

The .STL file is verified so that vertices match and triangles aren't missing, and then an egg-crate support structure is created with about a 0.2-inch cross-hatch. The support structure is used to anchor the part to the base during building, as well as provide a fixture for overhanging structures that would otherwise fall or float away before the build could complete.

The final .STL files, one with supports in addition to the original file, are then sliced into horizontal crosssections and saved as slice files. The slice files are then "merged" to create four separate files that control the SLA machine, ending with the file extensions L, R, V, and PRM. The largest, and most important, of these files is the V file, or *vector* file. This file contains the actual line data that the laser will follow to cure the shape of the part. Also important is the R, or *range* file, which contains the data for solid or open fills, as well as the recoater blade parameters.

The four build files are downloaded to the SLA, which begins building the supports with the platen at just above surface level. The first few support layers are actually cured into the perforations in the platen, thus providing a solid anchor for the rest of the part. By "building," the SLA uses the laser to scan the crosssection and fill across the surface of the resin, which is cured, or hardened into the cross-sectional shape. The platen is lowered as slices are completed so that more resin is available on the upper surface of the part to be cured. Figure 10.1 shows a schematic of the SLA process.

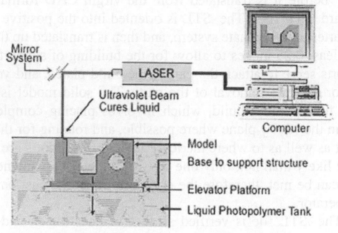

Figure 10.1 A simplified schematic of the SLA process.

Once a part is complete, the build platen is brought back up out of the vat of resin to its home position, which allows for excess resin to drain away. The final step is postprocessing, where the part must be removed mechanically from the base, and likewise the supports removed from the part.

10.3 Relation to Other Rapid Prototyping Technologies

The SLA systems currently provide probably the most accurate functional prototyping on the market. Although the postprocessing of SLA parts can sometimes be quite labor intensive, the smooth surface finish and high-dimensional tolerance acquired in SLA parts continue to advance even with newer systems. At this writing, there are about 1,000 SLA systems worldwide, and this is a number that continues to grow.

10.4 Applications of Stereolithography Parts

Other than purely functional concept models, SLA patterns can be used for investment casting and wind-tunnel modeling, as well as tooling. Foundries trained with using the SLA resin patterns can produce high-quality castings from them, and as seen in Chapter 18, the SLA provides an excellent alternative to machined models for some applications in the wind-tunnel. Finally, as discussed in Chapter 16, SLA pattern shells can be used as injection mold tools for limited runs. Figure 10.2 shows an SLA7000 system.

Figure 10.2 The SLA7000 is the latest line of stereolithography systems from 3D Systems.

10.5 Advantages and Disadvantages

Stereolithography offers a lot of advantage to a variety of business situations. SLA parts have probably the best surface quality of all other RP systems, and are also highly competitive in dimensional accuracy. Also, the latest SLA systems have significantly increased the speed at which parts can be produced, which is ultimately the goal of RP.

Finely detailed features, like thin vertical walls, sharp corners, and tall columns can be fabricated with ease even on older SLA systems, and the growing list of available resins are pushing the envelope on temperature and strength characteristics as well.

The main disadvantage of the SLA process is most likely the postprocessing requirements. Although significant advances have been made to make working materials safer and easier to work with, procedures to handle raw materials for the SLA still require careful and aware practices. Another disadvantage, which may decrease as resin competition increases, is the relatively high cost of photocurable resins, weighing in at around $600 to $800 per gallon.

10.6 Key Terms

Stereolithography. The first RP process, Stereolithography is a liquid-based RP system that cures epoxy resin with a low-powered laser to create three-dimensional models.

Photo-polymerization. The process of curing a plastic or polymer by introducing light as a catalyst.

Epoxy. The primary build material for sterelithography systems, epoxy thermo-set resins replaced the original acrylate resins.

Functional modeling. Refers to the rapid prototyping of directly usable parts, a capability for which Stereolithography is noted.

Casting patterns. Temporary patterns used to create a mold into which molten plastic or metals can be cast.

Injection mold tooling. Negative pattern halves used for large scale production of wax or plastic components. Injection mold tools must be able to withstand high stresses and temperatures as well as extensively repeated usage.

Please note that there are several extremely informative texts on the Stereolithography process, which is why this chapter was kept brief. See the recommended reading list at the end of this book for more comprehensive coverage of the SLA process.

For more information on stereolithography systems, contact 3D Systems at 805-295-5600.

11

Selective Laser Sintering

11.1 History of Selective Laser Sintering

The Selective Laser Sintering (SLS) process was developed by The University of Texas in Austin, and was commercialized by DTM, Corporation out of Austin, TX in 1987 with support from B.F. Goodrich. Since DTM is now essentially a subsidiary of B.F. Goodrich, the company has a strong parent to help absorb any financial burdens that may be incurred. The first SLS system was shipped in 1992, and there are currently several systems in use worldwide.

11.2 Selective Laser Sintering Technology

SLS is a rapid prototyping (RP) process that builds models from a wide variety of materials using an additive fabrication method. The build media for SLS comes in powder form, which is fused together by a powerful carbon dioxide laser to form the final product. SLS currently has 10 different build materials that can be used within the same machine for a wide variety of applications.

The SLS technology is housed in the Sinterstation line of systems by DTM. The current model is the Sinterstation 2500, which has various improvements over its predecessor, the Sinterstation 2000. Figure 11.1 shows the Sinterstation 2500.

The SLS process begins, like most other RP processes, with the standard .STL CAD file format, which is exported now by most 3D

CAD packages. The DTMView software can import one or several .STL files, and allows you to orient or scale the parts as you see necessary. The 2500 systems have "auto-nesting" capabilities, which will place multiple parts optimally in the build chamber for the best processing speed and results. Once the .STL files are placed and processing parameters are set, the models are built directly from the file.

Figure 11.1 The DTM Sinterstation 2500 (Courtesy of DTM Corp.).

The Sinterstations have a build piston in the center and a feed piston on either side. The models are built up in layers like other RP processes, so the build piston (15" x 13" x 16.7") will begin at the top of its range, and will lower in increments of the set layer size (0.003" through 0.012") as the parts are grown. With the build piston at the top, a thin layer of powder is spread across the build area by a roller/sweeper from one of the feed pistons. The laser then cures in a raster sweep motion the cross-sectional area of the parts being built. The part piston then lowers, more powder is deposited, and the process continues until all of the parts are built. Figure 11.2 shows the SLS system process chamber.

When the build media is removed from the machine, it is essentially a cake of powder with the parts nested inside. This cake is taken to the Break Out Station (BOS) table, where the excess powder is removed from the parts manually with brushes and hobby picks. The BOS has a built-in air handler unit to filter any airborne dust particles from the area, so no respiratory equipment is needed. Also, the BOS has a sock tube that attaches to a sieve on the table and to a powder canister underneath, so that the excess powder being removed from the parts can be kept for recycling and reuse.

At this point, some materials will require additional finishing, whereas others will be in end-use form. Some finishing techniques include glass-bead grit blasting (equipment can be purchased from DTM with the Sinterstation unit); sanding and polishing; drilling and tapping; and coating or infiltration.

Excess powder from each build for most materials can be recycled through a Vorti-Sieve for reuse in the system, therefore practically no material is wasted on support structures or the like.

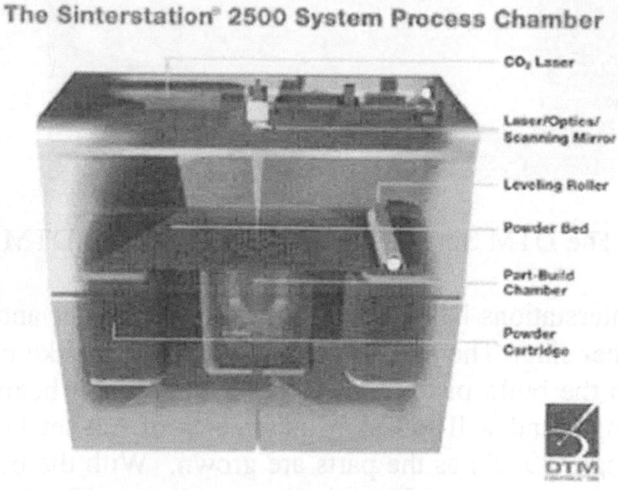

The Sinterstation® 2500 System Process Chamber

Figure 11.2 The SLS build chamber (Courtesy of DTM Corp.).

11.3 Purpose of Selective Laser Sintering

The SLS technology was developed, like other RP technologies, to provide a prototyping tool to decrease the time and cost of the design to product cycle. The strong point of the SLS process is that it can use a wide variety of materials to accommodate multiple applications throughout the manufacturing process. SLS was marketed early with three main applications: conceptual models, functional prototypes, and pattern masters. Since then they have added on an extra module, which incorporates rapid tooling.

Since the Sinterstation products are high end and require a large amount of up-front capital, the market range they targeted were large manufacturing industries with the capability to handle such specifications. DTM was looking to provide a cost-effective alternative for prototyping to these larger industries that spend millions of dollars to develop mass-produced products.

11.4 Current State of Selective Laser Sintering

The SLS technology currently has a high-quality product in the Sinterstation line, with their three main advantages being a wide range of build materials, high throughput capability, and the self-supporting build envelope. These advantages make the Sinterstation products better suited for industries with a wide range of needs and a demand for higher output. The main disadvantages lie in initial cost of system; peripherals and facility requirements; and maintenance and operation costs of the systems.

11.4.1 Advantages

11.4.1.1 Wide Range of Build Materials

The SLS technology currently employs 10 main build materials, which were previously grouped for sale into 3 central modules. Any or all of the build materials can now be purchased for use in the same Sinterstation machine, without requiring separate licenses. The modules are described as follows.

1. *The Casting Module.* The casting module include 5 different materials. All of the materials in the casting module are obviously directed at the metal casting/foundry industry, from investment shell casting to conventional sand casting. These materials are *Polycarbonate, TrueForm, CastForm,* and *SandForm Zr II & Si.*

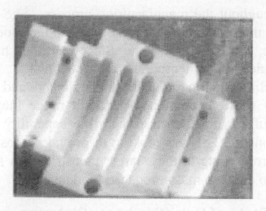

Figure 11.3 Polycarbonate material for the SLS system (Courtesy of DTM Corp.).

Polycarbonate was one of the original SLS casting pattern materials, but has now been discontinued from use. It is a fairly porous material, which allows for easy burnout from an investment shell for casting and a low 0.025% ash residue. The minimum feature size and wall thickness for polycarbonate parts is around 0.060", due to the brittle nature of the material and a 0.010" accuracy capability. The polycarbonate does run at a higher oxygen level, ~ 5.3%, therefore less nitrogen is used in keeping the build chamber inert. The porosity of the material also allows for the infiltration of epoxies or other thermosets to have stronger models, but some of the other materials are better suited for direct applications. Figure 11.3 shows a Polycarbonate part.

TrueForm is an acrylic-styrene polymer that was released after the polycarbonate material as casting pattern media. It has about a 1% ash residue on burnout, but can maintain a higher-dimensional accuracy, at 0.005", and can build thinner walls and features down to 0.030". The TrueForm has high feature and edge definition, and can therefore be used as secondary tooling patterns as well. Figure 11.4 shows a TrueForm part prior to removal from the SLS machine.

Figure 11.4 TrueForm material for the SLS system (Courtesy of DTM Corp.).

A new material in the casting module, and a successor to TrueForm, is called CastForm. The CastForm material is deemed to be more "foundry friendly" than even TrueForm, in that it requires less effort to burn out the pattern in the investment shell-firing process. Figure 11.5 shows a sample part fabricated from CastForm.

Figure 11.5 Sample casting pattern made with CastForm.

Figure 11.6 SandForm material for the SLS system (Courtesy of DTM Corp.).

SandForm Zr II and Si are direct Zircon and Silica foundry sands. They are for use as sand casting cores and molds and can maintain accuracy to 0.020". The sands can be used directly for casting, and provide the advantage of building complex cores that could not be produced using standard cope and drag techniques. Finally, the SandForm materials are compatible with both ferrous and aluminum casting processes. Figure 11.5 shows SandForm patterns.

2. *The Functional Prototyping Module.* The functional prototyping module consists of 5 different materials that are intended for direct-use applications as concept models, secondary tooling patterns, or functional hardware components. The materials licensed under the functional prototyping are *DuraForm, Nylon, Fine Nylon, ProtoForm,* and *Somos 201.*

DuraForm is a polyamide material recently released for creating highly detailed concept models. DuraForm has good surface quality, heat and chemical resistance, and can be polished for use in secondary tooling applications. Finally, it can be used to create features down to 0.030", and holds dimensional tolerances of 0.010". A DuraForm part is shown in Figure 11.7.

Figure 11.7 DuraForm material for the SLS system (Courtesy of DTM Corp.).

Nylon, Fine Nylon, and ProtoForm composite are the three nylon products used in the SLS process, that were essentially replaced by DuraForm. The nylons exhibit good toughness qualities, making them ideal for functional prototyping. The dimensional tolerance of the nylons is around 0.010", with a minimum feature size of 0.030". Finally, the ProtoForm composite, which is a glass-filled nylon, has strengths capable of withstanding high stresses in wind-tunnel type applications.

Somos 201 is a thermoplastic elastomer, marketed by DuPont, that has properties similar to some rubbers, which allows for flexible components (Shore A hardness = 81) to be directly rapid prototyped. This material is advantageous in applications such as seals, moldings and shoe soles where a functional flexible prototype is needed before expensive extrusion dies are created. The Somos 201 gets dimensional tolerances down to 0.010" and has elongation properties over 100%. Figure 11.8 shows a Somos 201 part.

Figure 11.8 Somos 201 material for the SLS system (Courtesy of DTM Corp.).

3. *The Rapid Tooling Module.* The rapid tooling module currently consists of three materials, which are *RapidSteel, Copper Polyamide, and LaserForm.* As more innovative direct application materials are introduced they will become part of this module.

Figure 11.9 RapidSteel material for the SLS system (Courtesy of DTM Corp.).

RapidSteel is a polymer-coated 1080 carbon steel powder that is fused in the SLS process to create a green part. This green part must then be fired in a furnace to remove the polymer binder, and the porous steel part is infiltrated, or wicked, with copper to produce the final metal component. The final product has strength and hardness properties much like aluminum, therefore it can be used to produce short-run tooling for preproduction plastic-injection molding or similar applications. The quoted tolerance is 0.010", before the fire and infiltration steps occur, wherewith after tolerances can range up to 0.030". Figure 11.9 shows a RapidSteel part.

Courtesy DTM, Corp.

Figure 11.10 Copper Polyamide tooling inserts made with SLS process.

Copper Polyamide is a polymer-coated copper than can produce directly usable soft tooling without the post-processing time and costs associated with RapidSteel. Unfortunately, strength and durability are sacrificed by going this route, so the application will ultimately choose with tooling material to use.

11.4.1.2 High Throughput Capability
The Sinterstation systems have high throughput capability compared to other RP machines due to several advantages. These

capabilities will vary between the different build materials, but overall can be described as follows.

1. *Scanning Speed.* The scanning speed is essentially the velocity of the laser movement across the part surface while it fuses the build material together. Since the systems are equipped with a powerful 50 watt carbon dioxide laser, the scanning speed for most of the build materials is very fast, so that large-part cross sections can be scanned in seconds. This high rate allows for multiple parts to be built in a short turnaround time.

2. *3D Part Nesting.* Since the SLS process builds parts in a powder-bed media, multiple parts can be "nested" throughout the build chamber in all axes. This allows the user to maximize the build output by completely filling the build chamber side to side and top to bottom with parts if necessary. This way the start-up and shut-down time is reduced as it is divided among many parts instead of just those that would fit in the −x, −y build plane. A second advantage to this system is that each part can be built with separate parameters, i.e. laser power, and if trouble occurs with one part in a batch it can be terminated without affecting the rest of the build.

3. *Large Build Envelope.* In conjunction with the high scan speed and three-dimensional nesting capabilities, a large build envelope (15" x 13" x 16.7") provides for a high part throughput in that many small parts or several larger parts can be fabricated in a single build run.

11.4.1.3 Self-Supporting Build Envelope

To cap off the exclusive advantage of the Sinterstation systems, a self-supporting build envelope provides several key bonuses over other RP technologies. Not only are parts faster to complete due to the lack of postprocessing, but they also stand less of a chance of being damaged during mechanical- or chemical-support removal. Also, as mentioned earlier, this self-supporting powder bed also eliminates any waste materials, so that only the material needed to create the part is used and the rest is recycled for building more parts.

11.4.2   Disadvantages

11.4.2.1 Initial Cost of the System
The initial cost of the Sinterstation systems range from $250,000 to $380,000, depending on the options and peripherals acquired and excluding facility modifications.

11.4.2.2 Peripherals and Facility Requirements
There are various peripherals necessary for optimum operation of the Sinterstation systems, including a BOS Table; air handler and sifter; and a glass-bead blaster for finishing. Combined, these components take up about 30 square feet of floor space. In addition, a Hydrogen Lindbergh furnace for firing and infiltration of the metal parts is necessary for the rapid-tooling module, which requires special facility and safety requirements for operating gases and general maintenance. All of these systems have various facility requirements in addition to the hard-wired 240V/70A power requirements of the Sinterstation itself, which also requires a large amount of floor space on the order of 200 square feet.   Finally, the Sinterstation weight combined is 6,275 pounds, which will require a sturdy floor to accommodate it.

11.4.2.3 Maintenance and Operation Costs
Since the Sinterstations are large and complex systems, the maintenance contracts currently run in the $35,000 annual range. Also, the powders must be properly stored and recycled for further use. The power consumption of the system and all its peripherals can be high and must be taken into account, along with the smaller costs of expendable inerting gases, build materials, and part finishing supplies.

11.5 Impact of the Technology

Although there are only a few hundred Sinterstation systems installed worldwide, the SLS process has had a significant impact on the prototype manufacturing industry. Mainly due to the wide range of build materials and higher output rates, the SLS systems have mainly been used in the RP job shop and production manufacturing arenas.

An impact directly on the RP industry was the release of the RapidSteel metal build material. Although it still has some drawbacks in that it is a multistep process that makes it much less "rapid," it was at least the first metal released for any RP system, which will more than likely lead to significant advances of this and other technologies in the future.

11.6 Interrelation with Other Technologies

The Sinterstation systems rank as one of the high-end RP systems due to the advantages listed previously. The part build times are on the average faster than many of the other systems, and thus the build times are used more effectively. The wide range of build materials is not seen in other RP systems as well, but there are disadvantages to deal with partially because of the advantages.

The capital investment for a Sinterstation 2500 system, as well as its large physical size and the extensive installation required are all characteristics not imposed by smaller RP systems. Also, the high cost of routine maintenance, expendable inerting gases used high electrical power usage can damper the effectiveness of the systems if they aren't used frequently and resourcefully. Otherwise, the network capabilities, ease of use and repeatability rank about the same as other RP systems used for similar applications.

11.7 Future of the Selective Laser Sintering Technology

The future of the SLS technology has high possibilities due to the robust systems designed from the start. With recent advances in powder metallurgy and ceramic powder technology, the SLS systems could be making quality ceramic and metallic hardware in the near future.

The first advancement will probably be in the metals realm, as DTM already has a head start with the RapidSteel system. Modifications to the RapidSteel, or alternative metallic powders that will result in fewer steps and more repeatability will provide a strong base for the future of SLS. The recent advent of LaserForm material may be the breakthrough the process is needing, as time will tell. Currently, several institutions and universities are vigorously working to evelop metal powders directly compatible with the SLS systems, and similar development is underway regarding ceramic powder sintering as well, so the horizon is looking pretty bright for the technology as a whole.

One interesting use of SLS in the future is the possible application of Sinterstations on lunar or Mars surfaces as manufacturing devices during and after colonization. Preliminary studies are underway with NASA using lunar simulant as a build material in a Sinterstation. The idea is that there would be an endless supply of build material (i.e., lunar dust) for fabricating parts as opposed to interplanetary shipping of building supplies.

11.8 System Update

Since the initial writing of this chapter, DTM Corp. has released the Sinterstation 2500*plus*, which has optimized operating parameters over the 2500, along with a Windows NT® software platform.

11.9 Key Terms

Self-supporting envelope. A unique feature of powder-bed RP systems, the workpiece does not require physical supports to be constructed for overhanging surfaces. The uncured powder-bed acts as the support material for such features.

Multiple build materials. A strong feature of SLS technology, many different materials can be used in the same machine without hardware modifications.

Three-dimensional part nesting. A bonus feature of powder-bed RP systems, parts can be placed in full three-dimensional space to optimize the per-part build time and postprocessing.

For more information on SLS contact DTM Corp at 512-339-2922.

12

Laser Engineered Net Shaping

Laser Engineered Net Shaping (LENS), is perhaps the first "true" direct-metal rapid prototyping (RP) system, in that parts are full strength metals upon removing them from the machine. Developed by Sandia National Laboratories and various industry members on a Cooperative Research and Development Agreement (CRADA), the LENS process uses virgin metal powders, per the user's preference, as build materials. The LENS 750 (12" x 12" x 12") and LENS 850 (18" x 18" x 42") systems are manufactured and sold by Optomec Design Company in Albuquerque, NM.

12.1 Build Materials

Current build materials with an extensive operational database on the system include Stainless Steel 316 (SS316), tooling steel (H13) and Titanium with 6% Aluminum and 4% Vanadium (Ti-6-4). Other metallic and ceramic materials have been tested and used at research facilities as well.

12.2 Build Process

Like most RP techniques, the LENS system uses a layered approach to manufacturing components, in which an STL file is sliced into horizontal cross sections, which are then downloaded to the machine from the bottom slice upwards.

12.2.1 Deposition Head

Metal powder is injected from 4 feeder tubes into the focal point of a high-powered laser, a 700W Nd:Yag in the case of the LENS 750, and the material is basically welded into place atop the previous layer. Figure 12.1 shows a schematic of the LENS process, whereas the actual building process is demonstrated in Figure 12.2. The system runs an inert atmosphere of argon to prevent oxidation of the powders during the build process.

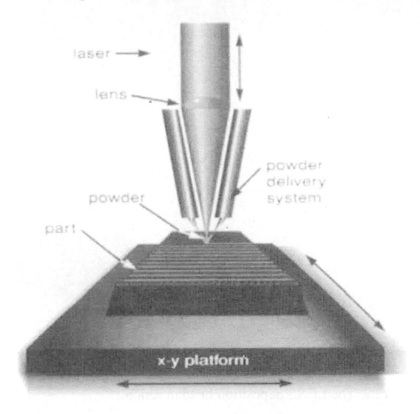

Figure 12.1 Schematic of the LENS process.

Figure 12.2 The LENS deposition head in action.

The material is deposited first as a perimeter of the current cross section, then a raster fill pattern is used to make the solid areas. There are 3 axes of motion, −x and −y provided by transverse movement of the deposition head, and −z provided by a moving platform either up or down. This provides the ability to produce simple to semicomplex three-dimensional objects, however objects with internal overhangs or free-hanging surfaces cannot be fabricated easily, due to rigid supports being required. The current maximum wall angle capability is approximately 18 degrees. However, the manufacturer is currently developing the software controls to allow for rotational, as well as a tilt capability for the deposition head. These enhancements will then allow for the fabrication of the most complex geometries, as they will negate the need for support structures.

12.2.2   Build Substrate

The build substrate is essentially a plate of the same material as the part which is being built upon it. The part is basically welded to this substrate, which prevents any movement or deformation during the build process. The part must be mechanically removed from the substrate once the part is completed.

12.3 System Statistics

Per specifications from the system vendor, the LENS system holds a dimensional accuracy of $\pm$ 0.020 inches, with a repeatability of about $\pm$ 0.005 inches in the $-x$, $-y$ plane and $\pm$ 0.020 inches in the $-z$ axis. Layer thickness can be varied from 0.001 inches to 0.040 inches. The deposition line width can be varied from 0.010 inches up to 0.100 inches, and the build rate is as high as 1.0 cubic inches per hour. The system requires a laser chilling unit, 208V/3-phase/100A power input, and about 80 square feet of floor space.

12.4 Postprocessing

First, as was mentioned earlier, the part must be cut from the build substrate. Current LENS parts tend to have a somewhat rough surface finish. Therefore, depending on the application, some machining may be required as well. No postsintering or curing is required, however, as the parts are full strength and density upon completion of the build.

12.5 Materials Properties

Tensile specimens were fabrication on Optomec's LENS system, and then compared against standard annealed bars of the same material for ultimate and yield strength, as well as elongation. The results were surprisingly in favor of the LENS system, regardless of the layered manufacturing technique. Table 12.1 shows the results of the tests, as reported by Optomec.

Build Material	Ultimate Strength (ksi)	Yield Strength (ksi)	Elongation (% per inch)
LENS 316 SS	115	72	50
Annealed 316 SS	85	35	50
LENS Inconel 625	135	84	38
Annealed Inconel 625	121	58	30
LENS Ti-6Al-4V	170	155	11
Annealed Ti-6Al-4V	130	120	10

Table 12.1 Materials properties of LENS-fabricated test specimens.

12.6 Typical Uses

LENS-fabricated components have the strength required to be used as end products, therefore direct metal prototypes may be used for product verification and testing or injection molding tooling.

One unique application of the LENS technology is the repair of existing metal hardware with the parent material. The low heat generated by the process doesn't have noticeable adverse affects on the original part. This opens up a new realm of hardware repair capabilities, where expensive hardware that is usually discarded when damaged may be refurbished for reuse. Figure 12.3 shows a sample LENS part with another part fabricated by a polymer RP process.

12.7 Advantages and Disadvantages

The key advantage of LENS is the capability to fabricate strong, functional metal hardware rapidly and directly from CAD data. Excellent materials properties as well as selection accent this advantage, as well as the fact that no post heat-treatment processes are required once the part is removed from the LENS machine.

The current disadvantage is a rough surface finish and low-dimensional accuracy acquired in LENS parts. Typically LENS parts must be polished or finish machined to fit required tolerances.

LENS™ produced part in 316 Stainless Steel

Stereolithography part

Figure 12.3 Sample LENS part next to a polymer RP component (Courtesy of Optomec Design Company).

For more information on LENS technology, contact Optomec Design Company at 505-761-8250.

13

3DP: Pro Metal System

Three-dimensional printing (3DP) is an MIT-licensed process, whereby liquid binder is jetted onto a powder media using ink-jets to "print" a physical part from CAD data. Extrude Hone Corporation incorporates the 3DP process into the Pro Metal system. The Pro Metal system is directed toward building injection tools and dies, as the powder used is steel based and durable enough to withstand high pressure and temperature. Models are built up from bottom to top with layers of the steel powder and a polymeric binder, printed in the shape of the cross sections of the part. The resulting "green" model is then sintered and infiltrated with bronze to give the part dexterity and full density. The Pro Metal system is one of only a few rapid prototyping (RP) processes to have metal material capability.

13.1 Pro Metal System Hardware

The Pro Metal system is currently available in only one size, the RTS-300, which can build models up to 12" x 12" x 12". The overall size of the modeler itself is approximately 8' x 5' x 7'. Parts built with the steel material are "green" and can be hardened with a few extra steps. The "green" steel part is placed in a furnace and sintered to remove the polymeric binder and bond the steel particles. This steel skeleton is then infiltrated with bronze in a second thermal cycle to completely densify the part.

The modeler has several important components, including the following.

1. <u>Build and feed pistons</u>. These pistons provide the build area and supply material for constructing parts. The build piston lowers as part layers are printed, while the feed piston raises to provide a layer-by-layer supply of new material. This provides the –z motion of the part build.
2. <u>Printer gantry</u>. The printer gantry provides the –x, –y motion of the part-building process. It houses the print head, the wiper for powder landscaping, and the layer drying unit.
3. <u>Powder overflow system</u>. The powder overflow system is an opening opposite the feed piston where excess powder scraped across the build piston is collected.

Extrude Hone Corp also sells a postprocessing package necessary for detail finishing and strengthening of the parts produced by the Pro Metal system. The package includes a firing/sintering furnace as well as various tools and expendables for cleaning and polishing the final components. Figure 13.1 shows the Pro Metal system.

Figure 13.1 The Extrude Hone Pro Metal RTS-300 3DP System (Courtesy of Extrude Hone Corp).

13.2 Pro Metal Operation

The Pro Metal has a very user-friendly interface, where only a very few commands are necessary to build a part. Since the parts are built in a powder bed, no support structures are necessary for over-hanging surfaces, unlike most other RP systems.

13.2.1 Software

The Pro Metal starts with the standard STL file format, which is imported into the Pro Metal software where it is sliced into thin, horizontal cross sections. When a file is first imported into the software, it is placed in an orientation with the shortest −z height. This is done as the fastest build capability, like other RP systems, is in the −x, −y direction. The part can be manually reoriented if necessary for best part appearance. Multiple STL files can be imported to build various parts at the same time for maximum efficiency.

Since the build envelop is a powder bed, three-dimensional nesting can be accomplished so that parts can be built in floating space to make room for others. This three-dimensional nesting capability is only available in a few other RP systems, and provides for a higher throughput of parts to be accomplished.

After the CAD file is imported and placed, a command is issued and the part file is sent to the machine to build. The part is built from bottom to top, slice by slice, until the complete model is made.

13.2.2 Machine Preparation for a Build

Before the part can be printed, the machine must be checked and ready. The feed piston should have sufficient powder added, and the build area is landscaped by the wiper blade until it is level with powder. The binder fill and takeup must be checked. Excess powder/dust around the printer gantry and throughout the chamber is vacuumed away for prolonged operation. Also, for optimum performance the moving parts of the system should be lubricated regularly.

13.3 Build Technique

The Pro Metal builds parts in a layer-by-layer fashion, like other RP systems. The following is a general description of a part build in progress.

1. The bottom cross section of the part is printed onto the first layer of powder. The jets print in the –y direction as the gantry moves in printer-head width increments in the –x direction.

2. When the layer is printed, the gantry remains at the left side of the table while the build-area piston lowers the slice-thickness amount.

3. The feed piston then raises a small amount, and the gantry sweeps across the part bed and overflow, spreading across a new layer of powder with a wiper. Excess powder is captured in the overflow.

4. The next layer is printed, and the process repeats until reaching the top layer of the part.

Figure 13.2 shows the previously described sequence of part-building steps in the Pro Metal system, with a bit more detail.

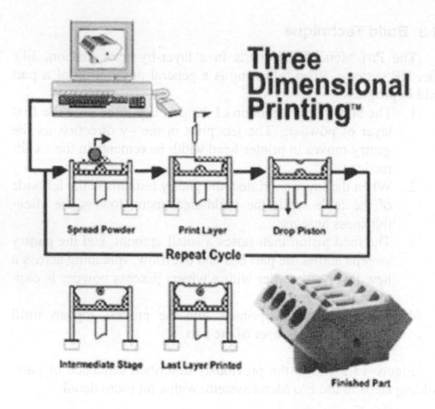

Figure 13.2 Pro Metal build sequence. Powder is spread across the build chamber. Binder is printed onto the powder in the cross-section shape. The build piston is lower and the feed piston raised. Repeat steps 1-3 until part is complete (Courtesy of Extrude Hone).

13.4 Postprocessing

Other than the Pro Metal system itself, there are several components needed for postprocessing of the part. For tooling or direct use, the parts must be sintered and infiltrated. Before infiltration, parts are fragile and must be handled with care. The following are the postprocessing steps for a part to be infiltrated with bronze, the typical infiltration metal.

1. <u>Powder Removal</u>. After the parts are taken from the machine, the excess powder must be removed. This is done

by brushing, vacuuming, and/or pouring out any unused powder from the parts.

2. <u>Thermal Cycle #1: Sintering</u>. Once the powder is removed from the part surfaces, the part is placed in an oven and heated to a temperature high enough to burn off the polymeric binder and fuse the steel particles into a 60% dense skeleton.

3. <u>Thermal Cycle #2: Infiltration</u>. The part is cycled in a furnace again, only this time bronze is melted and wicked into the steel skeleton, until a fully dense part is created (60% steel, 40% bronze).

4. <u>Finishing</u>. Depending on the application of the components, finishing can be done with conventional machining, polishing, and sanding techniques to the desired quality.

The actual postprocessing time will depend on the complexity of the part, the skill of the user, and the finishing technique used.

13.5 Typical Uses of Pro Metal

Parts built with the Pro Metal system are directly intended for use in a manufacturing environment. The steel-based material allows the molds to be used in conventional injection-molding presses, hence requiring no special adaptation to be used as a manufacturing tool. Some examples of typical uses of Pro Metal parts include injection molding, extrusion dies, direct metal parts, and blow molding patterns, as can be seen in Figures 13.3 through 13.6.

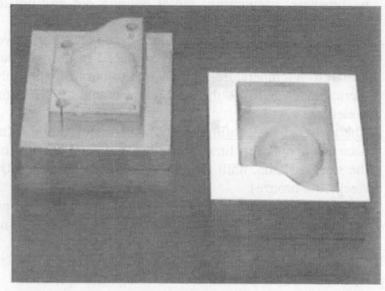

Figure 13.3 Injection molding tool fabricated with the Pro Metal system (Courtesy of Extrude Hone Corp.).

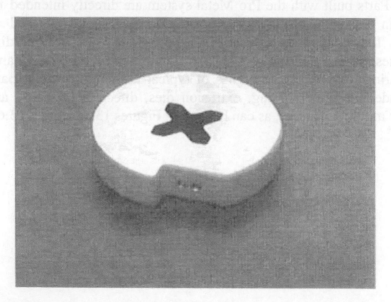

Figure 13.4 Extrusion die fabricated with the Pro Metal system (Courtesy of Extrude Hone Corp.).

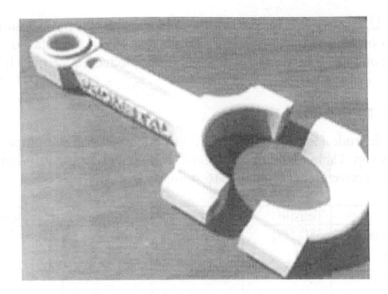

Figure 13.5 Direct metal connecting rod made with Pro Metal system (Courtesy of Extrude Hone Corp.).

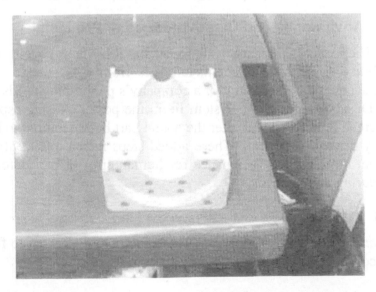

Figure 13.6 Blow molding pattern built with the Pro Metal system (Courtesy of Extrude Hone Corp.).

Ultimately, the speed and the material are the most desirable traits of the Pro Metal system. This is extremely advantageous to any company where time is a factor in preproduction or final fabrication of products.

13.6 Materials Properties

Extrude Hone Corp and MIT provided the material properties shown in Table 13.1 for Pro Metal components (post infiltration) versus conventional wrought 316 stainless steel.

Material	Tensile	Yield	Elongation	Hardness
Pro Metal	380-500 MPa	210 MPa	14-47%	60-63 HRB
Wrought 316	700 MPa	420 MPa	42%	89 HRB

Table 13.1 Materials properties of Pro Metal parts.

13.7 Advantages and Disadvantages of Pro Metal

The most significant advantage of the Pro Metal system is the capability to fabricate direct metal parts from CAD data. Thus, more avenues are opened up for a company's prototyping needs.

Disadvantages of the system lie in the postprocessing aspects, as there are multiple steps after the actual part build required to have a fully dense component. These added steps increase not only the product time but also the chance for human error or part deformation during thermal cycles.

13.8 Key Terms

Direct metal. The capability of a rapid prototyping system to fabricate components with metallic-based materials.

Green form. A soft, unfinished component that requires additional heat-treatment in order to achieve full strength and density.

Infiltration. Involves wicking a porous substrate with a lower temperature material to form a solid composite. In the case of the Pro Metal system, a low temperature metal is infiltrated into the porous steel model fabricated by the system.

For further information on the Pro Metal system contact Extrude Hone Corp at 724-863-5900.

14

Other Functional Rapid Prototyping Processes

Three other functional rapid prototyping (RP) processes, basically offered by service of the vendor, are covered briefly in this chapter. These processes, among other new processes, will probably continue to expand over the next few years as the need for directly functional prototypes increases.

14.1 Precision Optical Manufacturing

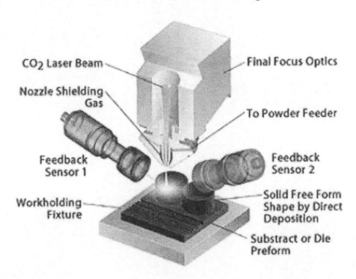

Figure 14.1 The precision optical manufacturing process.

Precision optical manufacturing (POM) is manufactured by the company of the same name, POM, and is a direct metals process not unlike the LENS technology. POM offers an electronic ordering system via the Internet where customers can upload files and have them fabricated and shipped directly to them.

POM focuses its energy on the rapid-tooling industry, building tool halves primarily with steels suitable for mass-production injection-molding runs. The POM process, shown in figure 14.1, deposits metal powder into the path of a very high-powered 2.5kW CO_2 laser beam, while the build platen is moved underneath to form the part shape.

The POM system has a closed-loop feedback system, where the melt pool of the deposited material is monitored and can be varied with software control to obtain optimum part quality.

In addition, the POM process contains an internal thermal post-curing process that serves to remove residual stresses left over from the build process.

As this process is new for the year 2000, advantages and disadvantages will be revealed as a user base becomes more widespread.

14.2 Laser Additive Manufacturing Process

The Laser additive manufacturing process (LAMP) is a large-scale manufacturing capability developed and operated at AeroMet Corporation in Eden Prairie, MN. The LAMP process uses a laser to sinter deposited titanium powder, much like the LENS and POM processes.

The key difference with LAMP is the high-deposition capability and large working envelope: the system deposits 10 to 12 pounds of metal per minute, at a 95% or better powder-usage rate, and has an expandable working area of a minimum 4' x 4' x 4'.

The laser system is very large, and the laser beam is actually piped in through a wall using fiber optic tubing. LAMP parts are typically built with about 0.050" tolerances, which are then final machined to meet dimensional requirements.

Figure 14.2 shows a schematic of the LAMP process.

The AeroMet™ Laser Additive Manufacturing Process

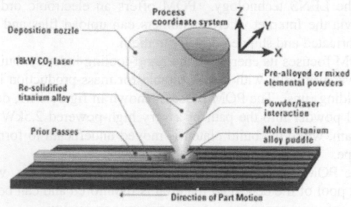

Figure 14.2 The LAMP process by AeroMet Corp.

14.3 Topographic Shell Fabrication

Topographic shell fabrication (TSF) is an RP service offered by Formus, Inc. in San Jose, CA. The TSF process builds large forms in a 10' x 10' x 10' working envelope. TSF builds parts in layers, by jetting paraffin wax into a sand base, with the final result being a wax/silica mix that can be finished and used as preforms for composite or fiberglass layups. Figure 14.3 shows the TSF process in action.

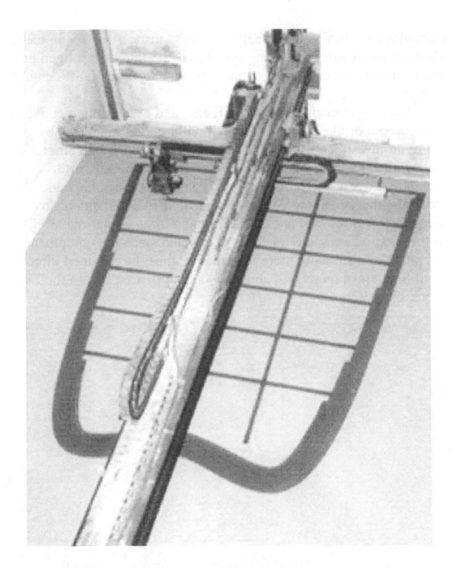

Figure 14.3 The TSF process building a large part.

14.4 Direct Shell Production

The direct shell production (DSP) process is based on MIT's Three Dimensional Printing (3DP) technology, with ceramic as the primary build material. Soligen, Inc, in Northridge, CA currently holds the license to the technology, and as opposed to other RP technologies they simply sell services instead of equipment. Through an electronic ordering system via the Internet, Soligen provides castings to customers, which are poured using DSP ceramic shells.

The customer provides an STL file, and Soligen then converts it into a tool of the actual part, which will be used in the same manner as an investment shell. The process works much like other 3DP technologies, where a binder is printed onto a powder surface, and then powder is added to the next layer and the process repeats. Afterwards, the remaining component is fired like a standard investment shell and the part is cast in aluminum or a similar metal. Due to the proprietary nature of the process, and since the equipment is not sold commercially, more information must be obtained directly from Soligen. Figure 14.4 shows a sample part cast from the DSP process.

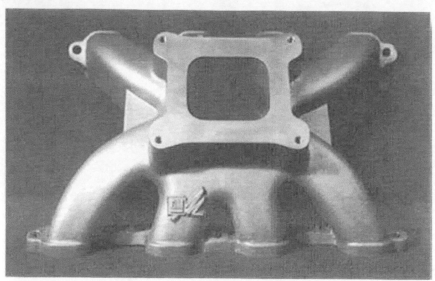

Figure 14.4 A sample part cast from the DSP process.

For more information on the DSP process, contact Soligen at 818-718-1221.

Unit III

Secondary Rapid Prototyping Applications

Secondary rapid prototyping (RP) applications are often required in order to obtain highly durable components from the RP processes. These applications include casting, tooling, and reverse engineering, which are covered in the following section. Also included are 2 case studies involving the direct application of RP in a real-world environment.

15

Casting Processes

15.1 Investment Casting

Since most of the rapid prototyping (RP) processes use a build material that either melts or burns away at high temperatures, rapid prototyped models can be directly used as investment-casting patterns. Investment casting is a highly used manufacturing process today, although it dates back to ancient origins. In the investment-casting process, an expendable model, or *investment,* is used to create a metallic part that can then be used as hardware.

This is done by first coating the investment with several layers of a ceramic shell material and sand. The pattern is then allowed to dry before it is placed into an oven or autoclave to burn or melt out the pattern material. The temperature is then raised to cure, or *fire,* the ceramic, to give it strength and higher resistance to thermal shock.

This hollowed ceramic cavity is now filled with molten metal, and once the metal has cooled the shell is broken away to reveal the usable metal part. The RP processes allow you to produce a prototype casting within a few days for minimal costs. This is a great advantage for a company trying to get a quick idea of how a product is going to perform, without going through the traditional and costly process of creating production tooling or machining. Figures 15.1 through 15.5 demonstrate the cycle of RP to investment casting.

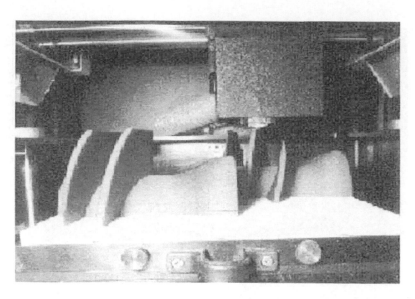

Figure 15.1 A model is rapid prototyped in wax to be used for investment casting.

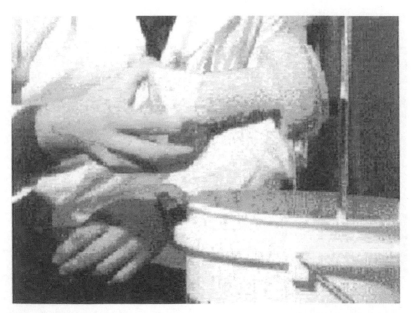

Figure 15.2 The wax model (investment) is shelled with layers of ceramic.

Figure 15.3 The ceramic shell is fired to remove the wax prototype and cure the ceramic.

Figure 15.4 Molten metal is poured into the evacuated shell and allowed to cool.

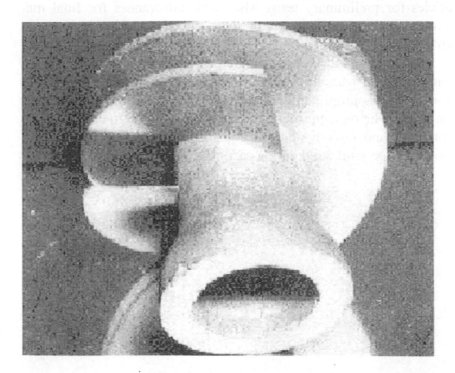

Figure 15.5 The ceramic shell is removed to reveal the final metal casting.

15.2 Sand Casting

Rapid prototyped models can also be used as patterns for *sand casting*. A sand pattern consists of sand mixed with bonding agents, which is all contained within a *flask*. In sand casting, a model is used to make an impression in two halves, or the *cope* and *drag* of the sand pattern. Molten metal is cast into this sand impression, allowed to cool, then pulled from the cavity and the process is repeated.

An RP pattern can sometimes withstand several applications of this technique before becoming distorted or damaged, therefore several castings can be made from the rapid prototyped model. Although a slight loss in dimensional accuracy may occur as compared to investment casting, this process can be used to create several test articles for preliminary tests. Also, with allowances for final machining in the model, it can produce directly usable hardware in small to large quantities.

The major drawback to sand casting is the limitations on part complexity. If parts have sharp edges, internal structures or protrusions with inappropriate draft, often times that geometry cannot be sand cast. The parts must be shaped so that the pattern can be pulled from the sand mold without disrupting the surface. Figure 15.6 shows a large sand mold during casting.

Figure 15.6 Large sand molds being cast with molten metal.

15.3 Permanent-mold Casting

Another common manufacturing technique, permanent-mold casting provides a means for higher production quantity as opposed to single-use investment shells. Permanent molds are machined of high-temperature material typically in two halves that are hinged together. When the halves are locked shut, molten metal is poured into the mold and allowed to cool. The mold is then opened to remove the cast part, and closed again to pour another one. In this respect, permanent-mold casting is analogous to plastic-injection molding, in that the tool is reusable many times to make identical parts.

16

Rapid Tooling

Rapid tooling refers to mold cavities that are either directly or indirectly fabricated using rapid prototyping (RP) techniques. *Soft tooling* can be used to inject multiple wax or plastic parts using conventional injection-molding techniques. Traditional *hard-tooling* patterns are fabricated by machining either tool steel or aluminum into the negative shape of the desired component. Steel tools are very expensive, yet typically last indefinitely, building millions of parts in a mass-production environment. Aluminum tools are less expensive than steel, although sometimes still $10,000 or more, and are used for lower production quantities up to several hundred thousand parts.

Soft tooling produces short-run production patterns (anywhere from 1 to 1,000 parts). Injected wax patterns can be used to produce castings as noted earlier, or injected plastic parts may be used directly in given applications. Soft tools can usually be fabricated for ten times less than a machined tool, as well.

RP can be used in this arena by several devices, from the direct production of tools on an RP machine to secondary, multiple-step processes.

16.1 Direct Rapid Prototyped Tooling

Rapid tooling is possible because some of the processes build with materials that are durable enough to withstand the pressures and temperatures associated with low-volume injection molding. Laser Engineered Net Shaping (LENS), Three Dimensional Printing

(3DP), and Selective Laser Sintering (SLS) all provide rapid proto-typed metal tooling capable of fabricating several thousand parts before tool failure.

In the case of SLS and 3DP, the components fabricated in the RP machine have to be postsintered and infiltrated with a lower temperature metal prior to being rigid enough to use. The LENS parts are directly usable strength-wise, however current technology doesn't provide adequate surface finish for most injection-molding requirements.

Stereolithography (SL) can also be used to fabricate short-run tooling as well. In this process, a thin epoxy shell of the tool is ac-tually built on the SL machine, and then is backfilled with a stronger, thermally conductive material. This technique allows for inlaying conformal cooling channels to help cool down the mold after each injection. SL tools have run up to 100 parts prior to fail-ure, but are typically used for quantities less than 50.

16.2 Silicone Rubber Tooling

Another route for soft tooling is to use the rapid prototyped model as a pattern for a silicone rubber mold, which can then in turn be injected several times. RTV silicones (Room Temperature Vul-canization) are preferable as they do not require special curing equipment. This rubber-molding technique yields a flexible mold that can be peeled away from more intricate patterns as opposed to firmer mold materials. There are as many or more techniques for silicone molding as there are RP processes, but the following is a general description for making simple two-piece molds.

First, an RP process is used to fabricate a positive image of the final component. Next, the pattern is fixtured into a holding cell or box and coated with a special release agent (often times a wax-based aerosol or a petroleum-jelly mixture) to prevent it from sticking to the silicone. The silicone rubber, typically in a two-part mix, is then blended, vacuumed to remove air pockets, and poured into the box around the pattern, until the pattern is completely encapsulated (this works best if clear silicone is used). After the rubber is fully cured, which usually takes 12 to 24 hours, the box is removed and the mold is cut in two (not necessarily in half) along a predetermined parting line.

At this point, the original pattern is pulled from the silicone mold, which can be placed back together and repeatedly filled with hot wax or plastic to fabricate multiple patterns. These tools are generally not injected due to the soft nature of the material, therefore the final part materials must be poured into the mold each cycle.

16.3 Investment-cast Tooling

Still yet another alternative is cast tooling. Metal shrinkage and other problems inherent to casting processes often prevent this from being a viable tooling technique, due to the high-accuracy required for injection-mold tools. The process is the same as investment casting the actual component, except that the tool is cast instead. The RP process is used to produce a model of a negative, which can be taken through a casting process to produce a metal mold, which in turn can survive many injections. Again this approach, unfortunately, allows for more dimensional error due to the many steps involved. It also causes an increase in turnaround time, which may be a determining factor in the selection process. But in some cases it can still prove more economically efficient than the traditional machining procedure.

16.4 Powder Metallurgy Tooling

Tools and inserts fabricated using powder metallurgy (P/M) provide long-life service comparable to machined tools, however they are made from rapid prototyped patterns. P/M tools are fabricated a few different ways, probably the most known, a process called 3DKeltool, is owned by the makers of stereolithography, 3D Systems, Corp.

In the P/M tool approach, a rapid-prototyped negative master is used to create a silicone-rubber positive. A proprietary metal mixture is then injected or poured around the positive, and is sintered to shape. Properties of P/M tools are similar to a tool steel, providing hundreds of thousands of parts prior to failure. P/M tools can usually be received within 2 to 4 weeks of production of the RP master, which remains competitive with a machined tool at current rates.

16.5 Spray Metal Tooling

Currently, several industrial and government groups are working to develop spray-metal tooling technologies. Thermal metal deposition technologies such as wire-arc spray and vacuum-plasma deposition are being developed to essentially coat low-temperature substrates with metallic materials. The payoff results in a range of low-cost tools that can provide varying degrees of durability under injection pressures.

The concept is to first deploy a high-temperature, high-hardness shell material to an RP positive, and then backfill the remainder of the tool shell with inexpensive low-strength, low-temperature materials and cooling channels (if necessary). This provides a hard, durable face that will endure the forces and temperatures of injection molding, and a soft backing that can be worked for optimal thermal conductivity and heat transfer from the mold. Although some successes are being achieved, the current stumbling block is the capability to deposit the harder, high-temperature material directly onto the RP pattern without affecting the integrity of the component geometry. One alternative is to use the RP pattern to create a silicone-rubber mold, which is used to create a ceramic spray substrate. This ceramic substrate can then endure the high-temperature metal spray. However, as it sounds, time and cost are multiplied by this approach.

In wire-arc spray, the metal to be deposited comes in filament form. Two filaments are fed into the device, one is positively charged and the other negatively charged, until they meet and create an electrical arc. This arc melts the metal filaments, while simultaneously a high-velocity gas flows through the arc zone and propels the atomized metal particles onto the RP pattern. The spray pattern is either controlled manually, analogous to spray painting, or automatically by robotic control. Metal can be applied, in successive thin coats, to very low-temperature RP patterns without deformation of the geometry. Current wire-arc spray technologies are limited to lower-temperature metals, however, as well as to metals available in filament form.

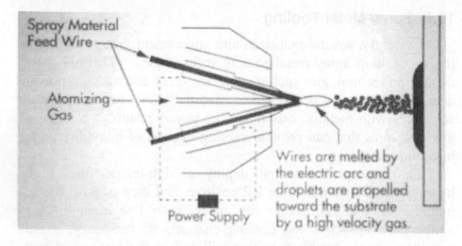

Spray Material Feed Wire

Atomizing Gas

Power Supply

Wires are melted by the electric arc and droplets are propelled toward the substrate by a high velocity gas.

Figure 16.1 The wire-arc spray process schematic.

Figure 16.2 Wire arc spray essentially "paints" metal onto a low-temperature surface.

Vacuum-plasma spray technologies are more suited to higher-melting temperature metals. The deposition material in this case comes in powder form, which is then melted, accelerated, and deposited by a plasma generated under vacuum. While vacuum-plasma spray is currently too hot for lower-temperature RP patterns, some limited success is being achieved with sturdier RP patterns such as SL or SLS.

16.6 Desktop Machining

Although RP is considered an alternative to machining, small, portable machining centers are now being produced for cutting molds and tools from softer materials. These systems use simple graphical user interfaces, which can import the STL data file, and can be set up and operated by novices in the machining area. Aluminum-based materials are milled out to quickly create short-run injection-mold halves without the need for numerical-control code programming or manual intervention. Some systems even offer cooling and tool-changing capabilities similar to larger machining centers.

17

Reverse Engineering Using Rapid Prototyping

Reverse engineering is a process used when an existing component must be reproduced, but the designs are not accessible. The part must be measured and redrawn to obtain new records, as well as to allow for the remanufacture of the component. This can be done several ways, from simply measuring the dimensions with calipers or micrometers and recording the data to be redrawn, to today's sophisticated techniques of translating x-ray imaging or three-dimensional laser scan data into computer-aided design (CAD) data files.

Reverse engineering, in light of rapid prototyping (RP), refers to the process of regenerating a physical object back into a digital CAD format, then producing direct or modified copies of the original object using RP. The application of reverse engineering spans a variety of industries, and can be accomplished with about as many techniques.

17.1 Process

Regardless of the environment in which it is applied, reverse engineering by RP follows a few basic steps. First, the physical object to be reproduced must be scanned electronically to provide a point cloud of coordinates along the various surfaces of the object. A point cloud is a set of coordinate data points in three-dimensional space representing a solid object.

This can be achieved by manual probes, which are placed against the object by hand and then the coordinate data at each point is entered into the computer by clicking a pedal or button. Coordinate measuring machines (CMM) are based on the same concept, except that the part is fixtured and the sensor tip is driven electronically to touch the surface and enter data points. Laser scanning provides data by passing a small laser or series of lasers across the surface and then receiving the coordinate data by reflection. One destructive technique removes layers from the part and photographically records each cross section, then converts the pictures into coordinate data slice by slice. Finally, x-ray, computed tomography (CT) or magnetic resonance imaging (MRI) scanning data may be used to create point clouds of a part's surfaces.

Secondly, once a point cloud is generated is must be converted from simple dots in space into actual surface representations. There are various "connect-the-dot" software applications available today that are designed for this specific purpose, which usually connect the points and export either the standard STL file for RP or a standard generic CAD file interface known as IGES (International Graphics Export Standard).

Finally, once the CAD or STL file is created, the parts can be rapid prototyped to provide the models or hardware as needed. Figure 17.1 shows a schematic of reverse engineering using CT scan data on a standard padlock.

17.2 Other Reverse Engineering Applications

Other applications of reverse engineering can be observed in the medical, electronics, and even archaeological fields. Surgery prototypes of internal organs, tumors, or bones may be fabricated from MRI data in order to prepare physicians prior to entering the operating room. Also, reverse engineering has been used to replicate dinosaur fossils prior to removing them from rocks, using scanned data. This was done in the event that if the actual article were to be destroyed during extraction from the rock, a replica could be fabricated from the computer data.

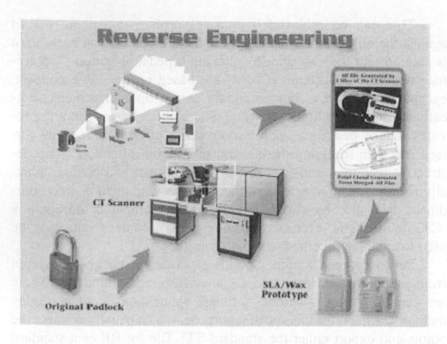

Figure 17.1 Schematic of the reverse-engineering process using CT scanning.

18

Case Study: Wind-tunnel Testing with Rapid Prototyped Models

As mentioned previously in this book, wind-tunnel testing is another venue where rapid prototyping (RP) proves its value. The machining of airfoils can be extremely costly and time consuming, and scheduling wind-tunnel testing time is very tight. Therefore, changes cannot typically be made during an experiment unless preliminary testing deems it necessary. With RP, the changes can be made essentially overnight, to provide for optimum design of products.

The following excerpt is from a study to test various RP models at multiple airspeeds against a machined baseline.

18.1 Abstract

Initial studies of the aerodynamic characteristics of proposed launch-vehicles can be made more accurately if lower-cost, high-fidelity aerodynamic models are available for wind-tunnel testing early in the design phase. This paper discusses the results of a study undertaken to determine if four RP methods using a variety of materials are suitable for the design and manufacturing of high-speed wind-tunnel models in direct-testing applications. It also gives an analysis of whether these materials and processes are of sufficient strength and fidelity to withstand the testing environment. In addition to test data, costs and turnaround times for the various models are given. Based on the results of this study, it can be concluded that RP models show promise in limited direct application for preliminary aerodynamic development studies at subsonic, transonic, and supersonic speeds.

18.2 Introduction

In a time when "better, faster, cheaper" are the words to live by, new technologies must be employed to live up to this axiom. In this spirit, a study was undertaken to determine the suitability of models constructed by RP methods for testing in subsonic, transonic, and supersonic wind-tunnels. This study was done to determine if the level of development in RP is adequate for constructing models using the current materials and processes, and if they can meet the structural requirements of subsonic, transonic, and supersonic testing while still having the high fidelity required to produce accurate aerodynamic data.

A study was undertaken to determine the feasibility of using models constructed from RP materials, using RP methods, for preliminary aerodynamic assessment of future launch-vehicle configurations. This study was designated to determine if the necessary criteria can be met satisfactorily in order to produce an adequate assessment of the vehicle's aerodynamic characteristics. The pertinent questions, or criteria, are as follows: (1) Can RP methods be used to produce a detailed scale model within required dimensional tolerances?; (2) Do the current RP materials have the mechanical characteristics, strength, and elongation properties required to survive wind-tunnel testing at subsonic, transonic, and supersonic speeds and still produce accurate data?; (3) Which RP process, or processes, and materials produce the best results?; (4) What steps and methods are required to convert an RP model to a wind-tunnel model (i.e., fitting a balance into an RP model and attaching the model parts together)?; and (5) What are the costs and time requirements for the various RP methods as compared to a standard machined-metal model?

This study was planned to answer the preceding questions by comparing RP models constructed using four methods and six materials to a standard machined-metal model. The RP processes used were fused deposition modeling (FDM), with materials of both ac-rylonitrile butadiene styrene (ABS) plastic and Poly Ether Ether Keytone (PEEK); stereolithography (SLA), with a photo-polymer resin of SL-5170 as the material; selective laser sintering (SLS), with glass-reinforced nylon as the material; and the laminated object manufacturing (LOM), using both plastic reinforced with glass fibers and paper as materials. Aluminum (Al) was chosen as the material for the machined-metal model. An aluminum model, while not as preferable as a steel model, costs less and requires less time to construct, thus providing a more conservative baseline model.

It can be stated initially that, at the current time, machined-metal models cannot be replaced by RP models for all required aspects of wind-tunnel testing. This study focused on a small aspect of wind-tunnel testing, that of determining the static stability aerodynamic characteristics of a vehicle relevant to preliminary vehicle configuration design.

While some RP methods and processes have reached a mature level of development—such as SLA, LOM using paper, and FDM using ABS plastic—others are still in the development phase or are trying new materials that promise greater material properties or higher-fidelity part definition. For the test, some of the materials and process still in the development phase were tested. This provided some models that did not meet visual standards and were not converted into wind-tunnel models. Two of these methods/materials were FDM using PEEK, and LOM using a plastic reinforced with glass fibers. FDM is now a standard RP technique, but using PEEK as a material is still in the early testing phases. PEEK provides models with much greater strengths but, at this time, the surface finish and tolerances on the models were unacceptable. LOM is a method that normally uses paper to construct a model, but this has low material properties, in other words, it is likely to break under the expected testing loads. A new LOM composite material is being tested to provide higher properties. At present, this material shrinks 3% during curing. From the model received, this 3% is not consistent over the model, since the model was warped and pitted. Due to these defects, this model was not converted to a wind-tunnel test article. The LOM paper model was converted into a wind-tunnel model, but the material delaminated during the machining of the bore hole and the installation of the balance adapter. The other three RP methods were wind-tunnel tested—SLA, SLS, and FDM-ABS.

18.3 Geometry

A wing-body-tail configuration launch-vehicle model was chosen for the actual study. First, this configuration would indicate whether possible deflections in the wings or tail due to loads, and whether the manufacturing accuracy of the airfoil sections would adversely affect the aerodynamic data that resulted during testing. Secondly, it would help answer the question of whether the model would be able to withstand the starting, stopping, and operating loads in a blow-down wind-tunnel. A photograph of the SLA wing-body model mounted in the transonic test section of the 14-inch Trisonic Wind-tunnel (TWT) is shown in Figure 18.1. The reference dimensions for the configuration are as follows:

Wing Body
$S_{ref} = 8.68$ in^2
$L_{ref} = 8.922$ in
XMRP = 6.2454 in (aft of nose)

18.4 Model Construction

The RP processes and materials selected for the baseline study were the following:

- FDM-ABS (by Stratasys) using ABS plastic, and FDM-PEEK using carbon fiber-reinforced PEEK
- SLA (by 3D Systems) using SL-5170
- SLS (by DTM Corp.) using glass-reinforced nylon (ProtoForm)
- LOM (by Helisys) using glass-reinforced plastic (LOMComposite) and wood (LOMPaper)

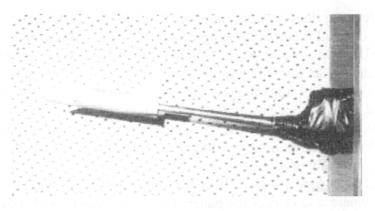

Figure 18.1 The wing body configuration prototype mounted in the wind-tunnel.

The RP models were constructed using the previously mentioned materials and processes, and are shown in Figure 18.2. These models were Al, FDM-ABS, SLA, and SLS.

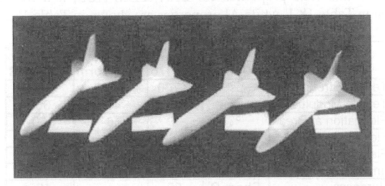

Figure 18.2 The four models tested left to right were Al, FDM-ABS, SLS, and SLA.

The FDM involves the layering of molten beaded ABS plastic material using a movable nozzle in 0.01-inch-thick layers. The ABS material is supplied in rolls of thin ABS filament resembling weed-trimmer line. The material is heated and extruded through a nozzle similar to that of a hot glue gun, which is deposited in rows and layered, forming the part from numerically controlled (NC) data. The PEEK material is currently being studied for the FDM process.

SLA uses a vat of photo-polymer epoxy resin which, when illuminated by an ultraviolet laser, solidifies the resin. The laser solidifies each layer, then the tray is lowered and the next layer is solidified. This continues until the part is complete.

SLS uses a laser to fuse, or sinter, powdered glass and nylon particles or granules in layers, which are fused on top of each other, as with the other processes.

LOM currently uses rolls of paper that are rolled onto the machine, where a laser cuts the pattern out of the sheet for that layer. The next sheet is rolled on top of the previous one, and also is cut. The sheets have epoxy on one side which, when heated, fuses the two sheets together. This is done by a hot roller after each sheet is cut, and the model is built up in this way. Currently, plastic is being tested as a material to replace the paper or wood, due to its better materials properties.

The materials properties for SLA, FDM-ABS, and SLS are shown in Table 18.1, while Al properties are shown in Table 18.2.

Property	Units	SLA SL5170	SLS Protoform	FDM ABS
Tensile Strength	psi	8,700	7,100	5,000
Tensile Modulus	ksi	575	408	360
Elongation at Break	%	12	6	50
Flexural Strength	psi	15,600	-	9,500
Flexural Modulus	ksi	429	625	380
Impact Strength	ft-lb/in	0.6	1.25	2
Hardness	(Shore D)	85	-	105

Table 18.1 Materials properties for various RP build materials.

Property	2024-T4	5086-H32
Yield Strength (ksi)	40	28
Tensile Strength (ksi)	62	40

Table 18.2 Strength properties for the Al model.

Each of the RP models was constructed as a single part. The nose section was separated from the core, and a 0.75" hole was drilled and reamed through the center of the body for placement of the aluminum balance adapter, which was then epoxied and pinned into place. The nose was attached to the core body using two screws that were attached through the nose to the balance adapter. An FDM model as built directly from the machine, is shown in Figure 18.3.

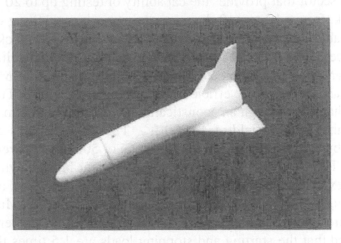

Figure 18.3 FDM model of winged body configuration.

18.5 Facility

The 14-inch TWT is an intermittent blow-down tunnel that operates by high-pressure air flowing from storage to either vacuum or atmospheric conditions. The transonic test section provides a Mach number range from 0.2 to 2.0. Mach numbers between 0.2 and 0.9 are obtained by using a controllable diffuser. The Mach range from 0.95 to 1.3 is achieved through the use of plenum suction and perforated walls. Each Mach number above 1.30 requires a specific set of two-dimensional contoured nozzle blocks.

A solid wall supersonic test section provides the entire range from 2.74 to 5.0, with one set of movable fixed-contour nozzle blocks. Air is supplied to a 6000 ft^3 storage tank at approximately minus 40°F dew point and 425 psig. The compressor is a three-stage reciprocating unit driven by a 1,500 horsepower motor. The tunnel flow is established and controlled with a servo-actuated gate valve. The controlled air flows through the valve diffuser into the stilling chamber and heat exchanger where the air temperature can be controlled from ambient to approximately 180°F. The air then passes through the test section, which contains the nozzle blocks and test region.

Downstream of the test section is a hydraulically controlled pitch sector that provides the capability of testing up to 20 angles-of-attack, from −10 to +10 degrees, during each run. Sting offsets are available for obtaining various maximum angles-of-attack up to 90 degrees. The diffuser section has moveable floor and ceiling panels that are the primary means of controlling the subsonic Mach numbers and permit more efficient running supersonically. Tunnel flow is exhausted through an acoustically damped tower to atmosphere or into the vacuum field of 42,000 ft^3. The vacuum tanks are evacuated by vacuum pumps driven by a total of 500 horsepower.

The 14-inch TWT, being an intermittent blow-down type tunnel, experiences large starting and stopping loads. This, along with the high dynamic pressures encountered through the Mach range, requires models that can stand up to these loads. It is generally assumed that the starting and stopping loads are 1.5 times the operating loads and are within the safety factor of 3 required for the wind-tunnel models. The worst starting loads occur at Mach 2.74, while the highest dynamic pressure of 11 psi is encountered at Mach 1.96. Table 18.3 lists the relation between Mach number, Reynolds number per foot, and dynamic pressure for the 14-inch TWT.

Mach Number	Reynolds Number	Dynamic Pressure
0.2	1.98 * 10^6/ft	0.60 lb/in^2
0.3	2.8	1.3
0.6	4.7	4.36
0.8	5.5	6.47
0.9	5.9	7.36
0.95	6.2	7.76
1.05	6.1	8.48
1.1	6.2	8.76
1.15	6.2	8.99
1.25	6.2	9.31
1.46	6	9.49
1.96	7.2	11
2.74	4.7	6.38
3.48	4.8	5.15
4.96	4.4	2.73

Table 18.3 Relationships of Mach and Reynolds number with dynamic pressure.

18.6 Test

A wind-tunnel test over a range of Mach numbers, from Mach 0.3 to 5.0, was undertaken in which each of the four models was tested to determine its aerodynamic characteristics. Three of the four models were constructed using RP methods, while the fourth acted as a control, being a standard machine-tooled metal model. A wing-body-tail launch-vehicle configuration was chosen as the test configuration. This was chosen to test the RP processes' ability to produce accurate airfoil sections and to determine the material property effects related to the bending of the wing and tail under loading. From a survey of past, current, and future launch-vehicle concepts, it was determined that a wing-body-tail configuration was typical for the majority of configurations that would be tested. The methods of model construction were analyzed to determine the applicability of the RP processes to the design of wind-tunnel models and the various RP methods compared to determine which, if any, of these processes would be best suited to produce a wind-tunnel model.

Testing was done over the Mach range of 0.3 to 5.0, at 13 selected Mach numbers. These Mach numbers were 0.3, 0.60, 0.80, 0.90, 0.95, 1.05, 1.10, 1.15, 1.20, 1.46, 2.74, 3.48, and 4.96. All models were tested at angle-of-attack ranges from −4 degrees to 16 degrees at zero sideslip and at angle-of-sideslip ranges from −8 to +8 degrees at 6 degrees angle-of-attack.

18.7 Results

For all phases of the baseline study, representative Mach numbers of 0.3, 1.05, and 3.48 are presented in this report. Coefficients of normal force, axial force, pitching moment, and lift over drag are shown at each of these Mach numbers. Only longitudinal data is shown for this study.

18.8 Baseline Models

The study showed that between Mach numbers of 0.3 to 1.2, the longitudinal aerodynamic data, or data in the pitch plane, showed very good agreement between the metal model and SLA model, up to about 12 degrees angle-of-attack, when it started to diverge due to assumed bending under higher loading. The initial SLS data for all the coefficients does not show an accurate representation of the process due to the model being a different configuration because of postprocessing problems. The second SLS model tested shows much better agreement with the data trends from the other models, but was not as good as the FDM and SLA. The greatest difference in the aerodynamic data between the models at Mach numbers of 0.3 to 1.2 was in total axial force. Between Mach numbers of 2.74 to 4.96, all the models showed good agreement in axial force. In general, it can be said that all the longitudinal aerodynamic data at subsonic Mach numbers show slight divergence at higher angle-of-attack due to the higher loading the models encountered; and at the supersonic Mach numbers, the data show good agreement over the angle-of-attack range tested.

18.9 Replacement Parts

Along with the baseline study, the replacement of standard ma-chined-metal model parts with those of RP parts was undertaken. The study showed that between Mach numbers of 0.3 to 1.2, the longitudinal aerodynamic data showed very good agreement be-tween the metal model and the metal model with the replacement FDM-ABS nose and SLA nose. The supersonic data show a slight divergence between the data, but the data trends are consistent.

18.10 Cost and Time

The RP models for this test cost between $3,000 and $3,500, and took between 2 to 3 weeks to construct, while the metal or alu-minum model cost about $15,000 and took 3.5 months to design and fabricate. RP fabrication costs for each model were between $1,000 and $1,500; conversion to a wind-tunnel model was about $2,000; and a balance adapter cost $100. The given costs are from quotes made by various secondary sources that specialize in RP part fabri-cation. It should be noted that the latest quote for the conversion of an RP model to a wind-tunnel model was $600 ($100 for the bal-ance adapter and $500 for parts and labor to convert the model). This was quoted as taking 2 work days. Along with the standard 3 days for RP model fabrication, a wind-tunnel model could be con-structed in under a week. The total cost for an RP model using in-house sources would be about $1,100, assuming $500 for RP mate-rial.

18.11 Accuracy and Uncertainty

The data accuracy results from this test can be divided into two sources of error or uncertainty: the model and the data-acquisition system. Each of these factors will be considered separately.

First, the dimensions of each model must be compared. Diffi-culty arose in the interface between the nose and core body for the RP models, along with the roll of the balance adapter in the models. Also, the contours of the models used in this test were measured at two wing sections, vehicle stations, tail sections, and the $-x$, $-y$ and

−x, −z planes. A comparison of model dimensions is shown in Table 18.4. This shows a representation of the maximum discrepancy in model dimensions relative to the baseline CAD model used to construct all the models at each given station. The standard model tolerance is 0.005".

The metal models balance adapter was rolled approximately 1 degree starboard wing down, while the RP models were rolled from 1 to 2.25 degrees starboard wing down, resulting in a difference of approximately 0 to 1.25 degrees between the two models (See Table 18.5). This results in a small error in all the coefficients, since the model was installed level in the tunnel. The installation of the balance adapter in both the metal and RP models was not a 0 degree roll. The effect of the balance adapters roll on the normal force and side force aerodynamic coefficients is shown in Table 18.6. Secondly, the repeatability of the data can be considered to be within the symbol size on the plots. Also, the capacity and accuracy for the balance used during this test is given in Table 18.7.

	Al	SLA	SLS	FDM	SLS 2
Wing L1	0.003	0.09	0.013	0.002	0.025
Wing L2	0.004	0.01	0.02	0.005	0.015
Wing R1	0.004	0.009	0.01	0.005	0.01
Wing R2	0.003	0.009	0.012	0.01	0.018
Body 1	0.01	0.002	0.002	0.02	0.007
Body 2	0.0005	0.002	0.012	0.006	0.015
Tail 1	0.002	0.002	0.014	0.001	0.009
Tail 2	0.001	0.002	0.014	0.002	0.015
XY Plane	0.001	0.015	0.016	0.006	0.014
XZ Plane	0.009	0.01	0.017	0.013	0.02

Table 18.4 Dimensional accuracy values at various segments of the models.

Model	Adapter Roll Angle
Al	0.95
SLA	2.25
SLS	1.05
FDM-ABS	1.57

Table 18.5 Roll angle variations of the models tested.

Balance Adapter Roll Effects		
	Cn	Cy
0.5 Roll	0.9999	0.0087
1.0 Roll	0.9998	0.0175
1.5 Roll	0.9997	0.0262
2.0 Roll	0.9994	0.0349
2.5 Roll	0.999	0.0436

Table 18.6 Balance adapter roll effects on the normal and side force coefficients.

	Strain Gauge Balance 250	
	Capacity	Accuracy
NF	200 lb	0.20 lb
SF	107 lb	0.50 lb
AF	75 lb	0.25 lb
PM	200 in-lb	0.20 in-lb
RM	50 in-lb	0.25 in-lb
YM	107 in-lb	0.50 in-lb

Table 18.7 Capacity and accuracy for Strain Gauge Balance 250 apparatus.

18.12 Conclusions

RP methods have been shown to be feasible in limited direct application to wind-tunnel testing for predicting preliminary aero-dynamic databases. Cost savings and model design/fabrication time reductions greater than a factor of 4 have been realized for RP techniques as compared to current standard model design/fabrication practices. This makes wind-tunnel testing more affordable for small programs with low budgets and for educational purposes. At this time, RP methods and materials can be used only for preliminary design studies and limited configurations due to the RP material properties that allow bending of model components under high loading conditions.

This test initially indicated that two of the RP methods were not mature enough to produce an adequate model. The methods were FDM using PEEK, and LOM using plastic. The LOMPaper model did not have high enough material properties to withstand the conversion process to a wind-tunnel model. The other three processes and materials produced satisfactory models, which were successfully tested. The initial SLS model did not produce good results due to problems with tolerances in postprocessing, but this was corrected in the second model and this process/material produced satisfactory results, although not as good as FDM or SLA.

FDM-ABS and SLA produced very good results for model replacement parts. The data resulting from the FDM-ABS model diverged at higher loading conditions, producing satisfactory results only for limited test conditions. It should be noted this material/process produced satisfactory results over the full range of test conditions for the vertical-lander configuration tested in a precursor study. SLA was shown to be the best RP process for wind-tunnel modeling, with satisfactory results for the majority of test conditions. The differences between the configurations data can be attributed to multiple factors such as surface finish, structural deflection, and tolerances on the fabrication of the models when they are "grown."

It can be concluded from this study that wind-tunnel models constructed using RP methods and materials can be used in subsonic, transonic, and supersonic wind-tunnel testing for initial baseline aerodynamic database development. The uncertainty, or accuracy, of the data is lower than that of a metal model due to surface finish and dimensional tolerances, but is quite accurate enough for this level of testing. The difference in the aerodynamic data between the metal and RP models' aerodynamics is acceptable for this level of preliminary design or phase A/B studies. The use of RP models will provide a rapid capability in the determination of the aerodynamic characteristics of preliminary designs over a large Mach range. This range covers the transonic regime, a regime in which analytical and empirical capabilities sometimes fall short.

However, at this time, replacing machined-metal models with RP models for detailed parametric aerodynamic and control surface effectiveness studies is not considered practical because of the high-configuration fidelity required and the loads that deflected control surfaces must withstand. The current plastic materials of RP models may not provide the structural integrity necessary for survival of thin section parts such as tip fins and control surfaces. Consequently, while this test validated that RP models can be used for obtaining preliminary aerodynamic databases, further investigations will be required to prove that RP models are adequate for detailed parametric aerodynamic studies that require deflected control surfaces and delicate or fragile fins.

18.13 Bibliography

Jacobs, Paul F., "Stereolithography and other RP&M Technologies," ASME Press, 1996.

Springer, A., and K. Cooper, "Application of Rapid Prototyping Methods to High Speed Wind-tunnel Testing," Proceedings of 86th Semiannual Meeting Supersonic Tunnel Association, October 1996.

Springer, A., K. Cooper, and F. Roberts, "Application of Rapid Prototyping Models to Transonic Wind-tunnel Testing," AIAA 97-0988, 35th Aerospace Sciences Meeting, January 1997.

Springer, A., "Application of Rapid Prototyping Models to High Speed Wind-tunnel Testing," NASA Technical Paper, 1997.

19

Case Study: Rapid Prototyping Applied to Investment Casting

19.1 Introduction

The objective of this project was to evaluate the capabilities of various rapid prototyping (RP) processes, and to produce quality test-hardware grade investment-casting models. The processes include the following:

- DTM Selective Laser Sintering (SLS),
- Stratasys Fused Deposition Modeling (FDM),
- 3D Systems StereoLithography (SLA),
- Helisys Laminated Object Manufacturing (LOM),
- Sanders Model Maker II (MM2), and
- Z-Corp 3D Printing (3DP).

19.2 Background/Approach

Investment-casting masters of a selected propulsion-hardware component, a fuel pump housing, were rapid prototyped on the several processes in-house. The models were then shelled in-house using a commercial-grade Zircon-based slurry and stucco technique. Next, the shelled models were fired and cast using a commonly used test-hardware metal. The cast models are compared by their surface finish and overall appearance (i.e., the occurrence of pitting, warpage, etc.), as well as dimensional accuracy.

19.3 Test Results

Eight copies of the fuel pump model in Figure 19.1 were slated to be fabricated, shelled, and cast by the end of this project. The following segment contains data for each of the models attempted, listed by the RP build process and initial pattern material. Data are summarized in the tables at the end of this section. Six of the eight models made it through the casting phase, whereas two did not. The part was not scaled in anticipation of casting shrinkage, which is reflected in the resulting dimensional data.

Explanations will be given within the proper segment for each resulting casting. Descriptive photographs of the different flaw terminology are shown in the Key Terms section. Dimensional data were taken in three axes with respect to the orientation of the part during rapid prototype pattern fabrication. Dimensions are given in inches over the entire part, with inch-per-inch accuracy in parenthesis.

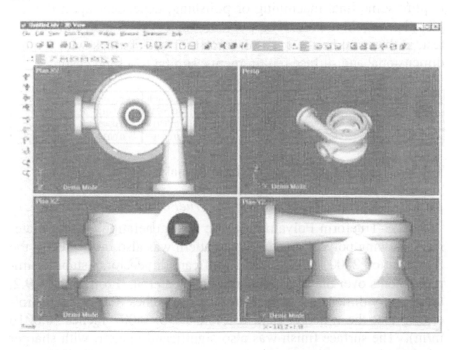

Figure 19.1 Various views of the baseline pump housing pattern.

19.3.1   Selective Laser Sintering - Polycarbonate Material

The SLS-Polycarbonate material was a precursor to the current preferred casting-pattern material of the SLS process, and was fabricated in the Sinterstation 2000 device. The casting-pattern fabricated showed a dimensional accuracy of approximately 0.010 inches (0.002 in/in) over the entire part in the –x and –y dimension, with significant accuracy in the z, or height dimension, of 0.003 inches (0.0008 in/in) (see Table 19.1). A significant amount of accuracy was lost from the pattern to the casting due to shrinkage via cooling of the metal casting. The maximum out-of-tolerance in the z dimension was 0.0529 inches (0.0124 in/in) over the entire part height.

The surface finish obtained had a grainy texture, due to the large standard particle size of the powder used for fabrication, which was picked up in the final metal casting as well. The surface finish of the final casting is about 200 micro-inches. There were no shell inclusions except for one chip in the lower thin rim. This finish may require some final machining or polishing, dependent upon the application requirements. The polycarbonate, per conclusion of this test, is therefore best suited for quick concept castings with lenient dimensional and surface requirements, similar to a sand casting.

SLS-Polycarb	x accuracy	y accuracy	z accuracy	x per inch	y per inch	z per inch
CAD-RP	-0.0105	-0.0090	0.0032	-0.0022	-0.0015	0.0008
RP-Casting	0.0439	0.0301	0.0496	0.0094	0.0049	0.0116
CAD-Casting	0.0334	0.0211	0.0529	0.0071	0.0034	0.0124

Table 19.1 SLS Polycarbonate dimensional analysis.

19.3.2   Selective Laser Sintering - TrueForm Polyamide

SLS-Trueform Polyamide is the next-generation casting material following polycarbonate. This pattern was also fabricated in the Sinterstation 2000, and showed significant dimensional stability improvements over the SLS polycarbonate material. From Table 19.2, the x and y dimensions were within 0.002 inches (0.0004 in/in) tolerance and the z height held a tolerance of 0.007 inches (0.0016 in/in). The surface finish was also significantly better, with sharper edges and a smoother texture than the polycarbonate. Shell burnout was much easier than with polycarbonate, with less ash content and a lower furnace cycle time. The final as-cast dimensional properties of the TrueForm casting were also excellent, with maximum out-of-tolerance in the z height dimension of 0.0155 inches (0.0036 in/in). The surface finish was near 60 micro-inches, with slight pitting due to shrinkage and ceramic slurry inclusions.

SLS-Trueform	x accuracy	y accuracy	z accuracy	x per inch	y per inch	z per inch
CAD-RP	-0.0020	-0.0010	-0.0070	-0.0004	-0.0002	-0.0016
RP-Casting	0.0104	0.0046	0.0225	0.0022	0.0008	0.0053
CAD-Casting	0.0084	0.0036	0.0155	0.0018	0.0006	0.0036

Table 19.2 SLS TrueForm Polyamide dimensional analysis.

19.3.3   Fused Deposition Modeling - Investment-casting Wax

For the FDM Investment-casting Wax, the vendor fabricated a model on an FDM-2000 at their facility to make up for the down time experienced during an on site equipment upgrade. The results herein are determined from that model (fabricated by the vendor of the FDM system), which was then shelled and cast on site in the same manner as the other specimens.

The wax pattern showed good dimensional tolerance, with a maximum discrepancy in the x dimension of 0.0077 inches (0.0017 in/in). The surface finish was very smooth, at 60 micro-inches, due to an apparent thin wax coating applied by the vendor. There were, however, a few minor inclusions and scabs due to ceramic slurry defects. The shell burnout was very simple and event-free, with no remaining ash content or shell disruption. Final as-cast dimensional properties revealed a maximum out-of-tolerance of 0.017 inches (0.004 in/in), which was measured in the z dimension of the part. The dimensions are shown in Table 19.3, and the pattern is shown in Figure 19.2.

FDM-Wax	x accuracy	y accuracy	z accuracy	x per inch	y per inch	z per inch
CAD-RP	0.0077	0.0003	0.0035	0.0017	0.0000	0.0008
RP-Casting	0.0006	0.0034	0.0135	0.0001	0.0006	0.0032
CAD-Casting	0.0084	0.0036	0.0170	0.0018	0.0006	0.0040

Table 19.3 FDM Investment-casting Wax dimensional analysis.

Figure 19.2 FDM wax pattern prior to shelling.

19.3.4   Laminated Object Manufacturing - High Performance Paper

The LOM High Performance Paper model (Figure 19.3) was fabricated on the LOM1015 machine.   The pattern dimensions showed a maximum tolerance discrepancy in the z dimension of 0.0293 inches (0.0069 in/in). There was a maximum variation from pattern to casting in the y dimension of 0.030 inches (0.0049 in/in), yet the shrinkage in the casting actually brought the final part back into close proximity of the original CAD data. The final as-cast dimensions revealed a maximum out-of-tolerance of only 0.0187 inches (0.0031 in/in).

The shell burnout did require extra attention for ash removal, although a light wax coating applied to the model prior to shelling reduced this significantly. Still, some cracking was noticeable in the shell after burnout, therefore the shell was redipped in the slurry in an attempt to patch it. This approach was partially successful, in that a final casting was achieved with a 60 micro-inch surface finish in most areas. There were "rat tails," however, on the casting due to metal seeping outward into some of the unfilled cracks in the shell. These can be removed by postmachining techniques, and really don't pose a problem. Table 19.4 shows the final dimensions.

LOM	x accuracy	y accuracy	z accuracy	x per inch	y per inch	z per inch
CAD-RP	0.0130	-0.0113	-0.0293	0.0028	-0.0018	-0.0069
RP-Casting	0.0003	0.0300	0.0286	0.0001	0.0049	0.0067
CAD-Casting	0.0132	0.0187	-0.0006	0.0028	0.0031	-0.0001

Table 19.4 LOM Paper dimensional analysis.

Figure 19.3 LOM pattern prior to shelling.

19.3.5   Three Dimensional Printing - Starch (Cellulose)

The 3DP pattern (Figure 19.4) was fabricated with the Z402 3D Printer system. Not unlike the other techniques, the major out-of-tolerance in the pattern was found in the z dimension at 0.0192 inches (0.0019 in/in). The as-cast dimensional analysis revealed the major out-of-tolerance, again in the z dimension, of 0.0543 inches (0.0127 in/in).

The 3DP pattern burnout was clean, similar to using a wax pattern. The final surface finish of the casting was about 300 micro-inches, with only a few surface inclusions and shrink pits. It was expected, due to the economical advantage of the 3DP process, that the castings would be somewhat rougher than more expensive processes, which turned out to be the case here. These patterns will provide faster, less expensive alternatives, however, to acquire near net-shape castings.

Z Corp	x accuracy	y accuracy	z accuracy	x per inch	y per inch	z per inch
CAD-RP	0.0065	-0.0085	0.0142	0.0014	-0.0014	0.0033
RP-Casting	0.0142	0.0349	0.0400	0.0030	0.0057	0.0094
CAD-Casting	0.0207	0.0264	0.0543	0.0044	0.0043	0.0127

Table 19.5 3DP Starch dimensional analysis.

Figure 19.4 3DP pattern prior to shelling.

19.3.6   Fused Deposition Modeling - ABS Plastic

The FDM-ABS (acrylonitrile butadiene styrene) model (Figure 19.5) was fabricated on the FDM1600 machine. Like the FDM Wax model, the ABS model held good tolerances with a maximum out-of-tolerance of 0.0068 inches (0.0016 in/in). The final casting had a maximum dimensional discrepancy in the x dimension of 0.0132 inches (0.0028 in/in).

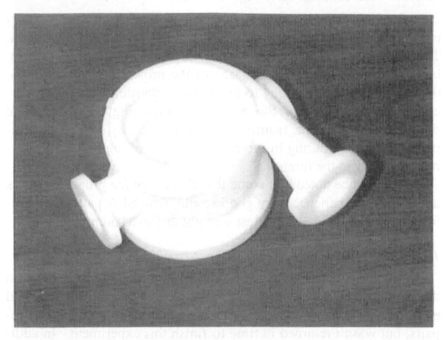

Figure 19.5 The FDM-ABS model as fabricated.

The surface finish suffered due to the pattern porosity, which allowed ceramic shell inclusion, sometimes completely through the thin planar sections of the component. Also, there was some shell cracking due to expansion of the material, which allowed some small "rat tails" in the casting where the molten metal flowed out into the cracks. Other than these problems, which could be corrected with pattern treatment, the majority of the surface finish was acceptable at 60 micro-inches.

FDM-ABS	x accuracy	y accuracy	z accuracy	x per inch	y per inch	z per inch
CAD-RP	0.0068	0.0003	0.0035	0.0015	0.0000	0.0008
RP-Casting	0.0055	0.0125	-0.0116	0.0012	0.0020	-0.0027
CAD-Casting	0.0132	0.0128	-0.0081	0.0028	0.0021	-0.0019

Table 19.6 FDM-ABS dimensional analysis.

19.3.7   Stereolithography - Epoxy 5170

An SLA epoxy pattern was made for this test with the vendor-recommended internal corrugated structure on the SLA-250 system. The pattern had good dimensional tolerance and surface finish, however, in the late casting phases of the project the pattern broke the shell during burnout due to excessive expansion. There are several possible causes for the expansion, including the incomplete drainage of extra resin from inside the part cavity, improper temperature ramping during burnout and the lack of high-oxygen burnout capability in the furnace.

It should be noted that some foundries with the proper facilities and training experience very good results with SLA patterns, therefore the model failure in this test was not due to a bad model.

19.3.8   Model Maker II - Investment-casting Resin

An investment-casting resin model was began fabrication on the Model Maker II system, but was discontinued due to repeated equipment failure. The machine was recalled for upgrades and repairs, but wasn't returned in time to finish this experiment. In addition, the model file was sent to the vendor for fabrication, which was again unsuccessful. The determining cause was concluded to be the excessive size of the part as compared to the capabilities of the equipment.

It should be noted that the Model Maker systems create very accurate casting models for much smaller size components than that which was used during this test.

19.4 Cost Comparisons

Process	Cost
SLS – Polycarb	$1,666.00
SLS – Trueform	$1,616.00
FDM – Wax	$1,765.00
FDM – ABS	$1,840.00
LOM	$2,005.00
Z Corp	$1,116.00
Sand Casting	$20,000.00
Machined From Plate	$65,000.00

Table 19.7 Summarized costs for each prototyping casting.

Table 19.7 shows the final costs estimated for each completed casting, compared against quoted figures for the same pattern fabricated by sand casting and machining. These figures included all associated expenditures from the point when the CAD file was received to the final castings and cleanup. A detailed table showing all of the variables concerned is attached as Appendix A.

19.5 Conclusions

Figure 19.6 shows the final as-cast patterns prior to gate-and-riser removal and post cleanup. Although the final-cast hardware components from each RP process were of varying degrees of success, each proved a significant cost advantage over conventional manufacturing techniques. The SLS-Trueform model provided the most acceptable casting, followed by FDM-Wax. Interestingly enough, these parts were in the intermediate cost range (Table 19.7), although the SLS pattern built 15 times faster than the FDM pattern (4 hours vs. 65 hours). This was due to a direct inverse proportion between the per-pound cost of the build materials and the cost of the machine run times.

Figure 19.6 As-cast hardware, prior to final cleanup.

The least expensive model was the Z Corp pattern, which also was the fastest to complete at 3.5 hours, although one of the least accurate. The Z Corp patterns will be more suitable for initial prototype castings, or especially castings that are designed for moderate final machining process (i.e., near-net shape castings).

Figure 19.7 Final hardware geometry: SLS TrueForm casting.

19.6 Key Terms

Slurry Inclusions. (Also referred to as *inclusions*). Slurry inclusions are rough patches, sometimes even holes, in the castings that are caused by the initial coats of slurry "wicking" into unsealed voids or porous material. These produce needle-like projections or bumps on the inside surface of the shell (postfired), that are then replicated as holes in the final casting. Figure 19.8 shows a typical slurry inclusion area, as seen on one of the castings in this study.

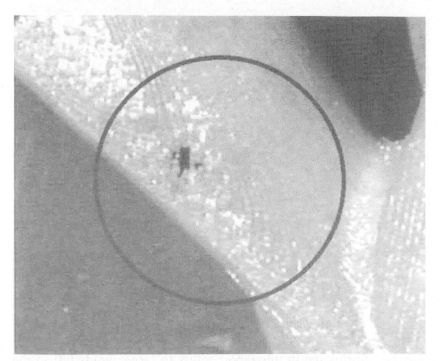

Figure 19.8 Slurry inclusion area on FDM-ABS casting.

Rat Tails. Rat tails are thin, protruding vanes of metal on the casting surface, that are caused by shell cracking during burnout. Shells crack for various reasons including excessive moisture buildup and over-expansion of the pattern. Although the shell sometimes will completely rupture due to cracking, more often it will just leave these small crevices throughout the shell. When the molten metal is poured into the shell during the casting process, it is allowed to flow out and fill up all of these cracks, thus resulting in protrusions on the final casting surface. Rat tails are much preferred over inclusions or shrinkage, as the excess material can be removed easily by finishing. Figure 19.9 shows an example of a rat tail.

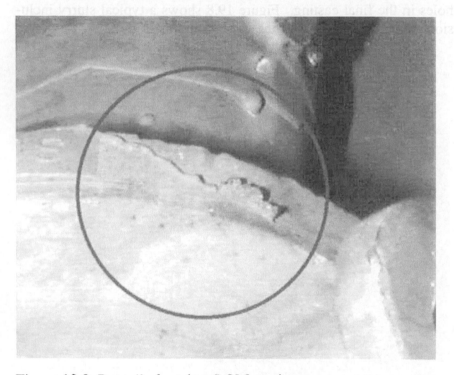

Figure 19.9 Rat tails found on LOM casting.

Shrinkage. Shrinkage in this report refers to larger voids left in the castings, often rendering the final components unusable. Shrinkage often occurs in thinner areas of a casting, which are adjacent to a dense section of the part. Although the complete area may be filled with the molten metal during casting, the cooling of the denser section occurs more slowly, thus "pulling" material away from the thinner areas. This is a common casting problem, which is confronted by allowing additional vents from strategic areas to give them more material to feed from during cooling. Figure 19.10 shows an example of shrinkage.

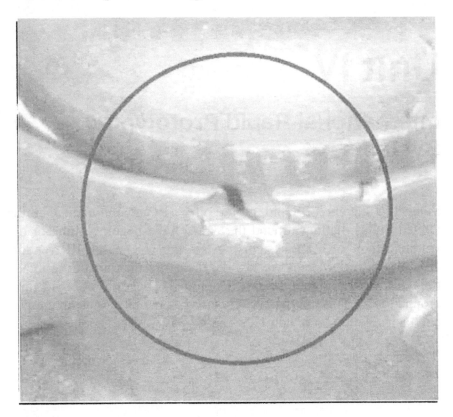

Figure 19.10 Shrinkage found on SLS casting.

19.7 References

Salvail, P., Cooper, K. "Proceedings of the Investment Casting Institute 47th Annual Technical Meeting," pages 3:1-3:6, San Francisco: Oct. 31 through Nov. 3, 1999.

Spada, A., ed. "Investment Casters Discuss RP...", Modern Casting magazine, pages 38-39, Vol. 90, No. 1, Jan. 2000.

Unit IV

International Rapid Prototyping Systems

Since the advent of rapid prototyping (RP) technologies in the United States, nations around the world have begun cloning or developing their own RP systems. Also, the U.S.-based RP vendors are now selling systems worldwide, therefore manufacturing industries around the globe are experiencing the benefits of rapid turnaround tools. This unit will briefly cover some of these technologies and their relation with their U.S. counterparts, where applicable.

20

RP Systems in Israel

20.1 Solid Ground Curing

Solid Ground Curing (SGC) is a resin-based rapid prototyping (RP) process manufactured by Cubital in Israel (see Figure 20.1). The process employs a photosensitive resin similar to stereolithography, in addition to a variety of other integrated subsystems to fabricate prototypes.

Figure 20.1 The SGC 5600 system by Cubital.

Gathering from the vendor's publicly available information, the process proceeds as follows: "A computer analyzes a CAD file and renders the object as a stack of "slices." The image of the working slice is "printed" on a glass photo-mask using an electro-static process similar to laser printing. That part of the "slice" representing solid material remains transparent. A thin layer of photo-reactive polymer is laid down on the work surface and spread evenly. An ultra-violet floodlight is projected through the photo-mask onto the newly-spread layer of liquid polymer. Exposed resin (corresponding to the solid cross-section of the part within that slice) polymerizes and hardens. The unaffected resin, still liquid, is vacuumed off. Liquid wax is spread across the work area, filling the cavities previously occupied by the un-exposed liquid polymer resin. A chilling plate hardens the wax. The entire layer, wax and polymer, is now solid. The layer is milled to the correct thickness. The process is repeated for the next slice, each layer adhering to the previous one, until the object is finished. The wax is removed by melting or rinsing, revealing the finished prototype. (Alternately, it can be left on for shipping or security purposes.)"

The SGC systems are physically quite large, set up in an automatic assembly line fashion with the curing lamp, resin vacuum, milling station, etc. They are currently offered in two systems, the entry level SGC 4600 and the high-end SGC 5600. The SGC 4600 systems offers a 14" x 14" x 14" working envelope, whereas the SGC 5600 system has a 20" x 14" x 20" envelope and a 300% increase in net production rate over the SGC 4600 system.

There are a few RP service bureaus successfully operating SGC system within the United States, as well as private corporations.

20.2 The Objet Quadra

Also from Israel, and new for the year 2000, Objet Geometries is set to release the Quadra system sometime later in the year. At this writing, information on the process was limited, but the following was offered by the system vendor (see Figure 20.2).

The Quadra process is based on state-of-the-art ink-jet printing technology. The printer, which uses 1536 nozzles, jets a proprietary photopolymer developed in-house by Objet. Because it requires no postcure or postprocessing, Quadra touts the fastest start-to-finish process of any (RP) machine currently on the market.

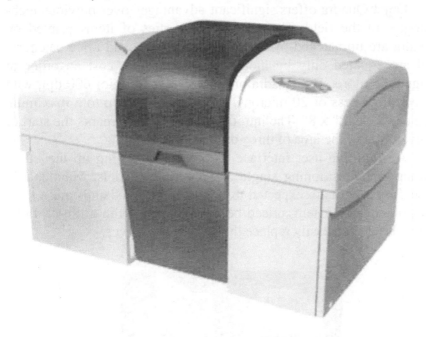

Figure 20.2 The Quadra system from Objet.

Objet will initially offer one grade of material with properties similar to multipurpose resins currently offered with competitive RP systems. Additional materials with varying properties are under development. Material is delivered by a sealed cartridge that is easily installed and replaced. Jetting of different resins, once they become available, will not require costly investments in materials or hardware upgrades. A new cartridge is dropped into place without any complicated procedures or specially trained staff.

Quadra deposits a second material that is jetted to support models containing complicated geometry, such as overhangs and undercuts. The support material is easily removed by hand after building the model. The support material separates easily from the model body without leaving any contact points or blemishes to the model. No special staff or training are required. Furthermore, models built on the system do not require sanding or smoothing where the supports are attached. Figure 20.3 shows parts from the Quadra.

Objet Quadra offers significant advantages over previous technologies in the field. The material properties of items printed on Quadra are unmatched by machines in its class and price category, and are equaled only by industrial systems that cost an order of magnitude more. The Quadra prints in a resolution of 600 dpi, with a layer thickness of 20 microns, and builds parts up to a maximum size of 11" x 12" x 8". The introduction of Quadra marks the start of a revolution in the area of three-dimensional imaging.

An intuitive user interface aids users in setting up the build, scaling, and positioning single and multiple models. Maintenance costs for Quadra are expected to be low. The UV lamps are a standard off-the-shelf item, priced below $75 each, with a life of 1,000 hours. Users can easily replace the lamps themselves.

Figure 20.3 Parts built with the Objet Quadra photocurable resin.

21

Rapid Prototyping Systems in Japan

21.1 Stereolithography

Japan has embraced the stereolithography technology in a big way, as there are many different types of liquid rapid prototyping (RP) systems available there. Table 21.1, courtesy of the Japanese Association of Rapid Prototyping, represents some of the companies with liquid RP systems, along with their build capabilities. European and American systems sold in Japan are also represented.

21.2 Laminated Object Manufacturing

Sheet-fed RP systems, in addition to the laminated object manufacturing (LOM) systems sold in the U.S. are also made and sold in Japan. Those systems are represented in Table 21.2, along with resellers of LOM systems.

21.3 Other Rapid Prototyping Systems

Most of the other RP systems available in Japan are imports from Europe and America, including FDM, SLS, and concept modelers. Also solid in Japan are the EOS systems from Germany. As can be seen, the manufacturing institutions of Japan have realized and capitalized upon the utility and advantages of RP to their design-to-manufacturing process.

RP Machine Name	Build Capacity (mm)	Vendor of System
SLA250Series50HR	254 x 254 x 254	incs inc. TEL: 813-5351-1919 http://www.incs.co.jp
SLA3500	254 x 254 x 254	
SLA5000	350 x 350 x 400	
SOLIFORM-250A	508 x 508 x 584	
SOLIFORM-250B	250 x 250 x 250	TEIJIN SEIKI CO,LDT TEL: 8144-813-8700 http://www.teijinseiki.co.jp
SOLIFORM-500B	250 x 250 x 250	
SOLIFORM-500C-S1	500 x 500 x 500	
SOUP250GH	500 x 500 x 500	
SOUP2 600GS	250 x 250 x 250	NTT DATA CMET INC. TEL: 813-3739-6611 http://www.nttd-cmet.co.jp
SOUP1000GS	600 x 600 x 500	
SOUP1000GA	1000 x 800 x 500	
SCS-300P	1000 x 800 x 500	
SCS-1000HD	300 x 300 x 270	DMEC LTD. TEL: 813-5565-6661 http://www.intio.or.jp/d-mec_mc/
JSC-2000	300 x 300 x 270	
JSC-3000	500 x 600 x 500	
SLP-4000R	1000 x 800 x 500	
SLP-5000	200 x 150 x 150	DENKEN ENGINEERING CO.LTD.
SOLIDER5600	220 x 200 x 225	TEL: 8197-583-5535 http://www.coara.or.jp/coara/dkslp/my.html
SOLIDER4600	508 x 356 x 508	KIRA Corporation
MEIKO(LC-510)	356 x 356 x 356	TEL: 81566-72-3171 http://www.kiracorp.co.jp
MEIKO(LC-315)	100 x 100 x 60	MEIKO INC.
UR2-SP1502	160 x 120 x 100	TEL: 81551-28-5111 http://www.yamanashi21.or.jp/xmeiko
E-DARTS	150 x 150 x 150	UNIRAPID INC. TEL: 81489-50-1460 http://webs.to/uri

Table 21.1 Liquid RP systems sold in Japan.

RP System Name	Build Capacity (mm)	Vendor Name
Solid Center (PLTA-A4)	280 x 190 x 200	**KIRA Corporation**
Solid Center (KCN-50N)	400 x 280 x 300	TEL: 81566-72-3171
		http://www.kiracorp.co.jp
LOM-1015E	370 x 240 x 360	**TOYODA MACHINE**
		WORKS,LTD.
LOM-2030H	810 x 560 x 490	TEL: 81566-25-5313
		http://www.toyoda-kouki.co.jp

Table 21.2 Sheet-fed RP systems (LOM) sold in Japan.

22

Rapid Prototyping Systems in Europe

22.1 Germany

Electro Optical Systems (EOS) in Munchen, Germany, develops and markets the EOSint line of systems, which are comparable to the Selective Laser Sintering (SLS) process in the U.S. There are three categories of EOSint line; the EOSint P, EOSint S, and EOSint M rapid prototyping (RP) systems, which build parts with polymers, foundry sand, and metals, respectively (see Table 22.1).

System	Build Capacity (mm)	Laser Type	Build Materials
EOSINT P 360	340 X 340 X 620	50W CO2	polystyrene polyamides
EOSINT M 250 X	250 X 250 X 185	200W CO2	steel
EOSINT S 700	720 X 380 X 400	2- 50W CO2	foundry sand

Table 22.1 EOS RP systems and specifications.

The EOS process uses a powder build media, which is sintered in layers using a laser. The powder is spread across the build platen in thin passes, while the laser melts the cross-sectional shape of the part onto the previous slice. The process is repeated until the complete part has been sintered. Unlike SLS, the metals are not in a polymer matrix, and hence need no further burnout or infiltration.

22.2 Sweden

Arcam, in Gothenburg, Sweden, is predicting to release its electron-beam direct-metal RP system in late 2000. The system will operate like other RP technologies, only with a pure metal powder. First, the .STL file is downloaded to the system software. Then, from a magazine of powder, an equally thin layer of powder is scraped onto a vertically adjustable surface. The first layer's geometry is then created through the layer of powder melting together at those points directed from the CAD file, with a computer-controlled electron beam. Thereafter, the building surface is lowered just as much as the layer of powder is thick, and the next layer of powder is placed on top of the previous. The procedure is then repeated so that the object from the CAD model is shaped, layer by layer by layer, until a finished metal part is complete (Figure 22.1).

Figure 22.1 Depiction of Arcam's electron beam RP process.

Arcam is also stating that the process will be environmentally safe and friendly enough to operate in an office environment! If this is the case, the design-to-manufacturing industry may be on the verge of some revolutionary changes, where actual parts will be made in minutes...right at the designer's station. This process will be closely watched by the RP community in the next few years.

22.3 Belgium

In Belgium, the KU-Leuven Production Engineering, Machine design and Automation group has posted the following information on a direct metals SLS-type system. Figure 22.2 shows the process.

"Selective Metalpowder Sintering (SMS) is a rapid prototyping technique which directly builds 3D metal parts out of metal powders. It consists of a 500 Watt CW Nd:YAG laser, an optical focusing unit, a XY-stage to move the beam over the powder surface, a powder recipient, a mechanism to deposit powder layers and a PC-based control unit. Different parameters influencing the quality of powder deposition have been investigated and optimized. This resulted in a slot feed mechanism followed by a cylinder to perform powder compaction. Minimum layer thicknesses of 0.1 mm have been obtained for Fe-Cu mixtures. The SMS set-up is now able to produce simple 3D metal parts by liquid phase sintering of Fe-Cu or Stainless Steel-Cu mixtures. Sintering has been done in an inert gas chamber to avoid contaminating oxidations. All samples are post-processed in a conventional sintering furnace. Future research aims at optimizing the process parameters in order to improve the product accuracy and strength and to accelerate the processing speed." Hopefully, more information will be available in the future.

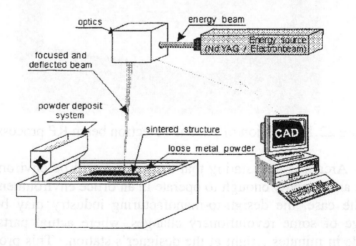

Figure 22.2 Schematic of the SMS process.

23

Rapid Prototyping Systems in China

The bulk of this chapter was extracted from the publicly published minutes to a previous meeting of the Rapid Forming Technology Committee (RFTC) of China. Brief descriptions are given to the rapid prototyping (RP) technologies being employed as well as developed in the People's Republic. The following is an acronym list required for navigation of this section.

RFTC Rapid Forming Technology Committee
TU   Tsinghua University
XJU   Xi'an Jiaotong University
HUST Huazhong University of Science and Technology
CLRF Center for Laser Rapid Forming (at Tsinghua U.)
NUAA Nanjing University of Aeronautics and Astronautics

23.1 Stereolithography

The Stereolithography method is studied by several universities, such as TU, XJU, HUST. Since 1992, the Stereolithography Apparatus (SLA) has been built up in the CLRF and TU. This is the first SLA equipment in China. XJU gave a systematic study on the SLA. According to their experiences, the most important problem is automatic stability of the liquid surface of photoresin. They improved the structure of their new RP machine, and got much better results.

23.2 Laminated Object Manufacturing

The Laminated Object Manufacturing (LOM) method is studied mostly in China. The CLRF, TU developed a new method-Slicing Solid Manufacturing, which is similar to LOM method, but has its own features. This new RP technology passed the inspection in 1993. The experts appraised that this new RP method obtained advanced level in the world.

23.3 Fused Deposition Modeling

The Fused Deposition Modeling (FDM) method is also studied by the CLRF, TU. They developed another new RP method-Melted Extrusion Manufacturing, which is similar to the FDM method, but has its own distinguish features. They are working on development of special nozzle design, filament fabrication of different materials, such as wax, nylon and ABS.

23.4 Selective Laser Sintering

The Selective Laser Sintering (SLS) method is also studied in China. A laboratory in NUAA did their effort to develop their new method. Their new RP machine has many different features, not only different with DTM, but also different with Long Yuan Co. In China. The AFS-300 RP machine made by Long Yuan Co. Was Sold to Tianjin and Beijing.

23.5 Multifunctional Rapid Prototyping Manufacturing System

The Multifunctional Rapid Prototyping Manufacturing System (M-RPMS) has been studied by the CLRF and TU for 6 years. The 3 New (New Concept, New Design, and New Products) is the feature of MRPMS-II machine. It puts together the two kinds of requirements in industrial applications and in research and development. The New Concept is the new designing principle, it is the most important guarantee for the developing success of the production. The New Design is used for the new technology of software, process, and equipment. The New Product is the M-RPMS-II machine. Each function provided by this machine can reach or exceed the quality of single-functional machines. Low operating costs and low operating time are the distinguished features of this machine.

Information for this chapter was gathered on the Internet at http://www.geocities.com/ CollegePark/Lab/8600/rftc.htm.

Appendices

Charts, Roadmaps, and Related Texts

These additions to the text will hopefully provide useful quick-referencing capability to the reader, for locating equipment vendors, checking the current state of the rapid prototyping (RP) industry and where it is going, as well as finding other RP-related texts for more in-depth study of specific areas.

Appendix A

Rapid Prototyping System Cross Reference Chart

The following chart provides specific information as to the names of the machines along with their domestic rapid prototyping (RP) system vendors, as well as to some of the system's basic specifications (see Table A.1).

Entries such as system cost and warranty would be ideal, however with the rapidly changing marketplace a static chart would simply not suffice. Typically, the office modeler systems are in the $30K to $60K range, whereas the functional *polymer* modelers are in the $125K to $300K range, and finally the direct metal systems are usually over $400K. Vendors offer varied warranties and maintenance contracts. Annual maintenance packages may cover anything from just the software up to a full "bumper-to-bumper" along with scheduled preventative maintenance visits by a technician to keep the system fine tuned.

The entries for speed are kept general due to part geometry and postprocessing issues as well as human intervention. For example, laminated object manufacturing (LOM) can build large, dense parts very quickly, however intricate parts can take a very long time to build as well as postprocess. Also, many processes (the laser-based systems) have the added bonus of economy-of-scale. In other words, by building many parts at the same time the average per-part time is significantly decreased.

System	Vendor	Process	Max Part Size (Inch)	Speed
JP System 5	Schroff Development	Knife-Cut Label Paper	24 x 24 x 6	Slow
Model Maker II	Sanders Prototype	Wax Ink Jet	6 x 12 x 8	Slow
ThermoJet	3D Systems	Wax Ink Jet	7 x 10 x 8	Very Fast
Z402	Z Corporation	Ink Jet Plaster Powder	8 x 10 x 8	Very Fast
Genisys	Stratasys	Polyester Extrusion	8 x 10 x 8	Fast
FDM 3000	Stratasys	ABS Plastic Extrusion	10 x 10 x 16	Moderate
Quantum	Stratasys	ABS Plastic Extrusion	24 x 20 x 24	Fast
LOM1015+	Helisys	Laser Cut Paper	10 x 15 x 14	Moderate
LOM2030H	Helisys	Laser Cut Paper	20 x 30 x 24	Moderate
SLA3500	3D Systems	Laser Cured Epoxy	14 x 14 x 16	Fast
SLA7000	3D Systems	Laser Cured Epoxy	20 x 20 x 23	Very Fast
Sinterstation 2500+	DTM Corp.	Laser Sintered Polymers	15 x 13 x 17	Fast
LENS 750	Optomec	Laser Sintered Metals	12 x 12 x 12	Fast
LENS 850	Optomec	Laser Sintered Metals	18 x 18 x 42	Fast
Pro Metal	Extrude Hone	Ink Jet Metal Powder	12 x 12 x 12	Fast

Table A.1 Selected RP machines, vendors, and specifications.

Appendix B

Direction of the Rapid Prototyping Industry

The following is a summary and current evaluation of "The Road to Manufacturing: 1998 Industrial Roadmap for the RP Industry," constructed by the National Center for Manufacturing Sciences Rapid Prototyping Technology Advancement Team. The Roadmap is now housed at the Georgia Tech Rapid Prototyping and Manufacturing Institute (RPMI), which has taken the roles and responsibilities for maintaining and updating it on a regular basis.

The Roadmap breaks up the various RP technologies into three distinct categories: 1) Design Verification Systems, 2) Bridge Technology Systems, and 3) Direct Manufacturing Systems. Granted, while some technologies can cross these boundaries into neighboring categories, they still serve as a good baseline to follow.

The Design Verification Systems (DVS), basically refer to the office modeler, or concept modelers as stated in this text. DVS are low-cost, low-accuracy, high-speed modelers aimed at the preliminary design phases of a project.

Bridge Technology Systems (BTS), are processes that provide a pattern for a secondary application, like investment casting or soft tooling.

Finally, Direct Manufacturing Systems (DMS), are technologies that meet the functional modeler criteria as laid out in this text. Basically, they provide working prototypes from the same or similar material from which the actual design will be fabricated.

The path from rapid prototyping (RP) to Advanced Rapid Prototyping (ARP), is identified as having three key elements that drives it: design technologies, materials and fabrication technologies, and integration technologies. These three pillars establish and support the whole of RP.

B.1 Design Technologies

Discussion goes through the processing power evolution of the personal computer from inception to as we know it. Details are listed as to the growth rate of processors, culminating to the currently approached doubling every 12 months. Interestingly enough, at the penning of the Roadmap, PCs were mentioned to have processing powers of 400 Mhz, whereas just recently both AMD and Intel released personal computers with 1.0 Ghz processors.

Also detailed is the progression of design systems from the pencil and slide rule up through the advanced solid modeling and Finite Element Analysis (FEA) packages today. "The driving function for future years is to enhance design systems to the point where complex designs can be created more quickly and by less-experienced users. Product generation could be accomplished or facilitated through automated techniques, including voice data entry, automatic solid creation from sketches, and 'smart' design systems with automatic filleting, stress analysis, and other automated functions."

B.2 Materials and Fabrication Technologies

This section of the roadmap relates to the core of this text; the actual RP hardware systems. Basically, predictions are made as to the development of the systems over the next decade, of which actually eight years are left. It branches into the three main RP system categories, which will be followed here as well.

B.2.1 Design Verification Systems

This branch suggested that for DVS to continue to advance, they must see an order-of-magnitude reduction in the cost of the

systems, maintenance, and build materials, as well as similar increases in quality.

As a follow-up to these requests only 2 years later, a few examples of progress already come to light. While the costs haven't
changed much, the quality is definitely seeing some improvement.
First, the Z402 systems have released much stronger materials that
allow also for smooth finishing in their fast concept modeler. Also,
the ThermoJet system was released, which jumped a great hurdle in
reliability and performance experienced by its predecessor, the Actua 2100 system. Finally, releasing in late 2000 is the Quadra system (see Chapter 20), which has over 1500 ink jets and actually
builds parts with a photocurable resin similar to SLA. If successfully implemented as planned, the system will provide fast, accurate
models at a concept modeler price, but with properties comparable
to those achieved by the BTS.

B.2.2 Bridge Technology Systems

The BTS mentioned, including laminated object manufacturing
(LOM), selective laser sintering (SLS), stereolithography apparatus
(SLA) and fused deposition modeling (FDM), were basically fore-
seen as undergoing only small changes, as they are seen as reaching
a plateau in their development. While their utility will still be in-
creasingly sought, the actual advancement of these technologies
may not go much farther.

As predicted, some of the systems have seen some improvements, with the release of the FDM3000 and SLA7000 systems,
providing increases in speed as well as materials properties. Also,
SLS has realized better casting and functional prototyping materials
for their Sinterstation systems, as well as a reduction in cost below
the $300K threshold.

B.2.3 Direct Manufacturing Systems

Some of the most exciting advancements have occurred in just
the past two years for the DMS. While the roadmap called for DMS
that can create nonorthogonal laminates, functional gradients, and
selective alloying, several systems have already been or are cur-

rently being released that produce fully functional metal components.

The laser engineered net shaping (LENS) process is offering functional grading capabilities with direct metals, while precision optical manufacturing (POM) and laser additive manufacturing process (LAMP) are providing high speed production of tool steel and titanium parts, respectively. Also, soon to be released by Arcam is an electron-beam direct-metal RP system, touted to be safe enough to operate in an office environment. The advent of all these systems are pushing full steam ahead toward the goals set by the Roadmap, perhaps even a bit sooner than expected.

B.3 Integration Technologies

The third and final tier of the Roadmap discusses advancement in Integration Technologies, which are grouped into several categories including standards, libraries, and collaborative infrastructure, in addition to reverse engineering and human-machine interface. Updates will be provided here on the first three, which have seen the most significant advances over the past 2 years.

B.3.1 Standards

A call is made for more stringent standards solely dedicated to the RP industry. As additive manufacturing processes form objects quite differently, the past databases of standards for machined or cast parts can not easily be applied. This is a situation that continues to be debated and will hopefully soon be addressed. Several committees have been formed by various industry RP professionals and technical personnel, which meet at selected RP conferences to discuss and debate the ideal properties standards for RP. Perhaps soon, conclusions will begin being drawn on to optimum patterns that may be used and recognized internationally for properties such as surface finish, tensile strength, and dimensional accuracy.

B.3.2 Libraries

This segment suggests virtual libraries that collect RP knowledge for access by individual users, as well as vendor-specific catalogs to document product, process, and materials information. With the help of the Internet and international electronic mailing lists, these requirements should be able to easily come to fruition over the next few years.

B.3.3 Collaborative Infrastructure

The path here states that "manufacturing users have the underlying need to access and control information across distributed computing systems for specific applications." One stated example is a virtual system that allows a user to upload a design and instantly receive a quote for production, along with suggestions for a better design and even provide an electronic verification of the client's credit line before executing the job. Among other examples, just recently, QuickParts.com debuted as a fully virtual RP service bureau, which allows for the user to upload their CAD file, select their preferred RP process, receive an instant quote, and then submit or decline the job in a previously established electronic account.

While a significant amount of networking and development with the design and CAD industry will have to be undergone before the system can "recommend alternative suggestions to the design for optimal performance," the groundwork has already been laid with the virtual RP service bureaus.

B.4 Summary

The RP industry appears to have adopted and is very much following this Roadmap significantly. In addition to the exerts presented here, there is a wealth of information and interesting predictions and suggestions stored in the document, along with very helpful graphics and data tables for expected changes in the RP arena. A highly recommended document for any persons interested in the future of the manufacturing industry as we know it.

Special thanks to the architects of this roadmap, the RPTA, who poured their time and energy into drafting such a solid document. The Road to Manufacturing: 1998 Industrial Roadmap for the Rapid Prototyping Industry" can be purchased directly from the NCMS.

Appendix C:
Recommended Rapid Prototyping Publications

The following is a list of books and publications dedicated to the rapid prototyping (RP) industry, some of which have served as a basis for university coursework and the like. Most were not used for reference in this text, and although I have read some of them and would highly recommend them, others I haven't, so good luck in your queries!

Jacobs, Paul F., "Stereolithography and other RP&M Technologies, From Rapid Prototyping to Rapid Tooling," ASME Press .

Kochan, D., ed., "Solid Freeform Manufacturing: Advanced Rapid Prototyping," New York: Elsevier Science Publishers, 1993.

Beaman, J.J., J. W. Barlow, D.L. Bourell, R.H. Crawford, H. L. Marcus, K.P. McAlea, "Solid Freeform Fabrication: A New Direction In Manufacturing," Boston, Kluwer Academic Press, 1997.

RPA/SME, "Rapid Prototyping Systems: Fast Track to Product Realization," SME, 1994.

Jacobs, Paul., "Rapid Prototyping and Manufacturing: Fundamentals of Stereolithography", SME.

Johnson, J.L., "Principles of Computer Automated Fabrication," Palatino Press Incorporated.

Burns, Marshall., "Automated Fabrication" Prentice Hall .

Wood, Lamont., "Rapid Automated Systems: An Introduction", New York, Industrial Press Inc., 1993.

Bennett, Graham., "Developments in Rapid Prototyping and Tooling," London, Mechanical Engineering Publications Limited.

Index

T - #0050 - 111024 - C0 - 229/152/14 - PB - 9780367397654 - Gloss Lamination